STATE MACHINES IN VHDL Dividers Vol. 3

State Machine Design for Arithmetic Processes

Daryl Ray Hawkins

Copyright © 2014 Daryl Ray Hawkins
All rights reserved.
ISBN: 1494351358
ISBN-13: 978-1494351359

Revision Dates
January 2015 First Release

Table of Contents

Contents

- Overview 4
- 1 Prerequisites 4
- 2 Fixed Versus Floating-Point 5
- 3 Limiting Factors 7
  - 3.1 Carry propagation 7
  - 3.2 Integer Size 7
- 4 Normalizing, Rounding, and Bounds 9
  - 4.1 Fixed-Point 10
  - 4.2 Floating-Point 12
  - 4.3 Rounding 14
  - 4.4 The Sticky Bit 15
  - 4.5 Divide By Zero 16
- 5 Simple Sequential Division 17
- 6 Signed Sequential Division 26
  - 6.1 Signed Sequential Using Conventional Wisdom 27
  - 6.2 Signed Sequential Using On-The-Fly-Conversion 44
- 7 SRT Division 61
  - 7.1 SRT Basics 61
  - 7.2 SRT Radix-2 Division 65
    - 7.2.1 Architecture Overview 66
    - 7.2.2 Leading Digit Normalizer 70
    - 7.2.3 SRT Algorithm's Principal Data Paths 76
    - 7.2.4 Carry-Save Form 82
    - 7.2.5 Signed Digit Representation 82
    - 7.2.6 Fast Carry-Select Adders 84
    - 7.2.7 Denormalizer 86
    - 7.2.8 Correction, Rounding and Sticky Decode 86
- 8 Other information on Dividers 146
- 9 Addendum 147

Overview

Compared to all other arithmetic functions, implementing a divider in hardware is the most demanding. A divider is more difficult to design and requires more logic resources. More clock cycles, hence a longer execution time. Consequently, multiplication based algorithms are often used in place of dedicated hardware divider, even within the microprocessor world.

Even so, there are several practical solutions available to the designer, as provided in this book. While speed, resource use, and power consumption are typically competing priorities, resource use is more demanding with division, as the reader will see.

Copies of all source code, in text file form, used in this book may be acquired through the web site listed below -- under publications, or pubs.

http://www.hawkinseng.com

1 Prerequisites

State Machine techniques used throughout this book are covered in the "STATE MACHINES IN VHDL *Composition* Vol. 1" book, which is a prerequisite, and should be reviewed by the reader.

Additionally, the "STATE MACHINES IN VHDL *Multipliers* Vol. 2" book contains useful information on fast adders and rounding techniques.

Finally, reviewing the IEEE 754-1985/2008 specifications is also recommended.

2 Fixed Versus Floating-Point

All mathematical operations are inherently fixed-point and are favored over floating-point, primarily because of the additional overhead burden associated with floating-point normalization -- pre and post. The down side to fixed-point is reduced accuracy when approaching the lower boundaries, and the magnitude confinements of its explicit range.

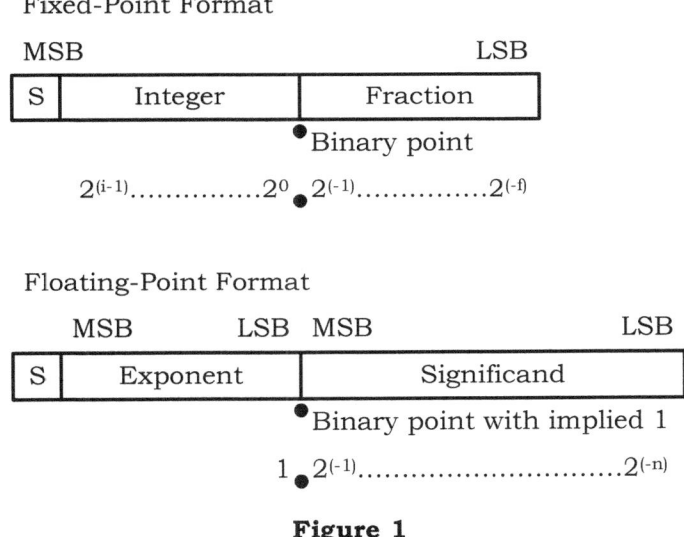

Figure 1

Fixed-point is a two's complement binary number with a pre-defined number of integer and fractional bits with the binary point being implicit between the two. The industry Q notation for fixed-point binary numbers is Qm.n, 'm' representing the number of integer bits excluding the sign bit; 'n' representing the number of fractional bits.

Note: *An example for Qm.n would read Q15.16 for a 32-bit value.*

Floating-point is a signed magnitude number with an implied binary point. The significand (all bits to the right of the binary point) is left-shifted so that the most significant '1' bit is positioned to the left of the implied binary point and discarded (not saved), while adjusting the exponent (bias) accordingly.

Two of the most common implementations are single and double precision, 23-bit and 52-bit significand (mantissa) excluding the implied 1; 24-bit and 53-bit including it.

The number of bits allocated for the exponent is 8-bits for single precision and 11-bits for double. Their nominal value is an offset from 0, which is +127 and +1023, respectively. Because they are offset values instead of absolute values, they are referred to as a bias instead of an exponent; sometimes called an exponent bias.

Associated extended formats are sometimes used for division but are not needed with multiplication. Extended means that more bits are used during the operational stage than are kept after normalization.

> Note: *Chapter 4 provides a conceptual overview of how division is done for both fixed-point and floating-point, and what bits are used for rouding. However, all implementations in this book are done in fixed-point. Where warranted, notes are provided on floating-point for consideration by the designer.*

3 Limiting Factors

3.1 Carry propagation

An identical section on carry propagation is also found in "STATE MACHINES IN VHDL *Multipliers* Vol. 2", which provides direction on creating and utilizing fast adders to achieve higher performance. This information also applies to division and should be reviewed by the reader.

The use of limited or carry-free adders are introduced in later chapters this book, along with digit-set representation. Digit-sets, in higher forms, enable carry information to be embedded or inferred within each digit, thus avoiding the use of traditional adders and their long carry propagation chains.

There are many forms of digit-sets. Details regarding their use and types are provided in the chapters and sections in which they are used.

3.2 Integer Size

There is no inherent limit to the size of an adder in VHDL. There are, however, limitations in representing and monitoring large fixed-point numbers in source code.

Generally, designers use conversion functions to convert between real-number data objects and binary numbers. Real signals or variables can be expressed textually in the floating-point style. This is useful when assigning constants in the source code and for monitoring objects during simulation.

The exponent (**) operator and CONV_INTEGER (std_logic_arith.vhd) or TO_INTEGER (numeric_std.vhd) functions are limited a range between plus and minus 2147483647 (2147483648), reducing the conversion range of the integer portion to 32-bits (sign included) and the fractional portion to 31-bits (no sign), each.

> Note: *Provided in the addendum are functions that will convert from fixed-point to real-numbers (simulation floating-point data*

objects) that extend past the integer limitation of VHDL. They may also be used to assign constants with floating-point text representation in synthesizable source code.

4 Normalizing, Rounding, and Bounds

Unlike multiplication where the resulting product contains all the needed bits for normalizing, rounding, and overflow detection; division requires that all quotient bits be generated individually by re-evaluating the remainder after each cycle.

The number of cycles needed to produce the integer portion of the quotient is equal to the number of bits of the operands: divisor and dividend. Additional cycles are required to generate fractional bits, and even more for those bits used in rounding. The first statement assumes both operands are the same number of bits.

Overflow occurs when the magnitude of the quotient exceeds the bounds of the resulting operand, normalized or not. A zero data value can qualify as an underflow.

Rounding involves generating at least two extra bits, which are used in conjunction with the last remainder value. This closely mimics the guard, round, and sticky or GRS paradigm used in multiplication.

> Note: Refer to "STATE MACHINES IN VHDL *Multipliers* Vol. 2" sections 4.3 and 4.4 for additional information on rounding bits and modes.

4.1 Fixed-Point

The biggest advantage to using fixed-point is that the binary point is aligned between operands with no exponent. Just as with the pencil and paper method, the binary point is moved or virtually repositioned to the right of the divisor's LSB, as it is with the divided, thus dictating the binary point position of the initial quotient.

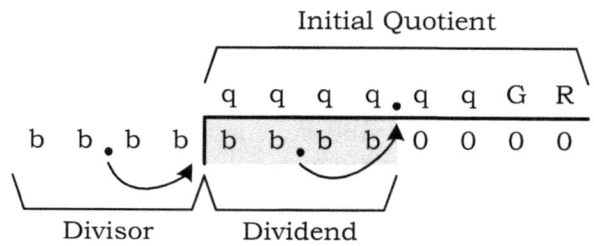

Figure 2

The number of cycles equal to the operand size generates the integer portion of the quotient. The number of cycles equal to the fractional portion of the operand plus two rounding bits, generates the fractional bits plus the guard (G) and round (R) bits, as shown in Figure 2.

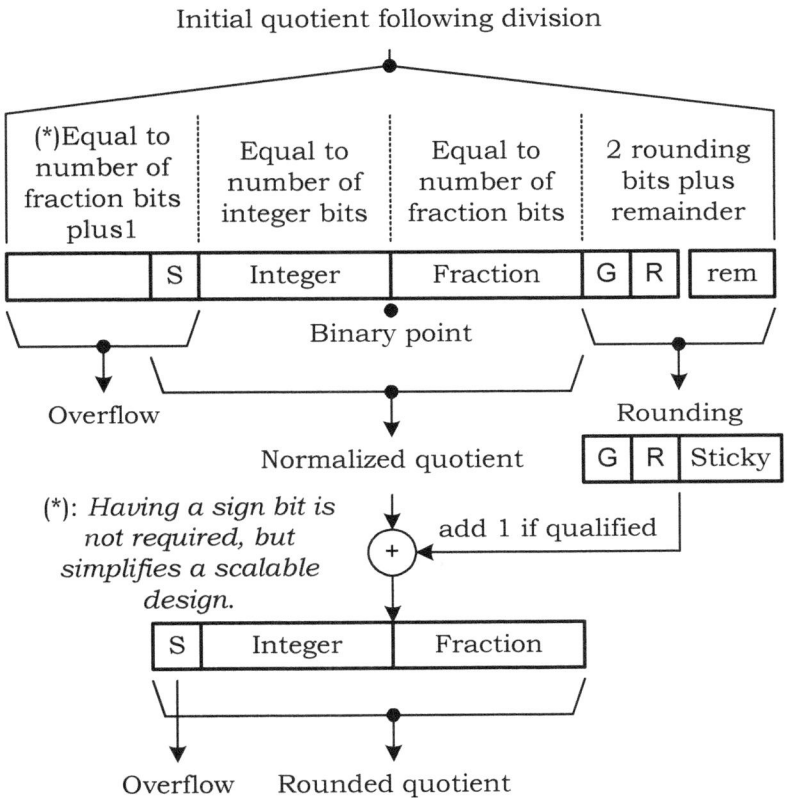

Figure 3 Fixed-Point Normalizing

Figure 3 shows the steps for finalizing the quotient following a fixed-point divide operation. The binary point position of the normalized quotient is offset from the right by the number of fraction bits plus two. Bits to the left of the binary point represent the integer portion, sign, and additional bits used for overflow detection.

1). Initial quotient size = 1 + Integer bits + (2 x Fraction bits) + 2
2). Normalized binary-point position = Fraction bits + 2

If the last remainder value is used to generate the value of the sticky bit, see section *4.4 The Sticky Bit*. Combined with bit G and R, determine whether the normalized value is rounded or not.

Note: *the smaller the number of fraction bits used in the implementation, the greater the round up error can be.*

If the sign bit of the normalize quotient is not equal to the additional overflow bits immediately following the division, an overflow has occurred. If the data portion, including both the integer and fraction fields, are zero after rounding, this constitutes an underflow, except if the result is a negative number. This last point is because, if the MSB is a one and is followed by all zeros, the number is the maxium negative number supported by that format.

4.2 Floating-Point

Unlike fixed-point, the binary point need not be virtually repositioned, as shown in Figure 4. Instead, the implied 1 is reasserted to the left of each significand, as is, and the exponent of the divisor is subtracted from the exponent of the dividend while maintaining the relative bias. The bias value is equal to 127 for single precision and 1023 for double.

Initial quotient exponent = (divided EXP – divisor EXP) + bias

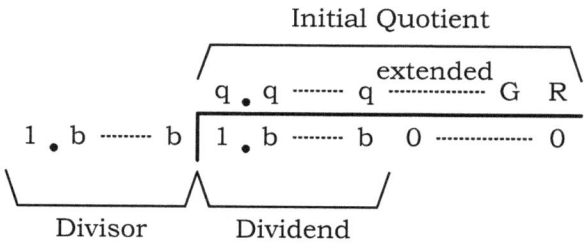

Figure 4

The number of cycles is equal to the precision, either 24-bits for single and 53-bits for double, plus extended bits, plus two rounding bits. IEEE stipulates 8 or more extended bits for single precision and 11 or more for double for extended formats, totaling 32+ and 64+, respectively. Some white papers suggest the number of extended bits be equal to the precision itself, resulting in an initial quotient of (2 x 24)+2 and (2 x 53)+2.

The additional bits improve the accuracy of the final product after rounding, or prevents truncation if the initial quotient is a very small value.

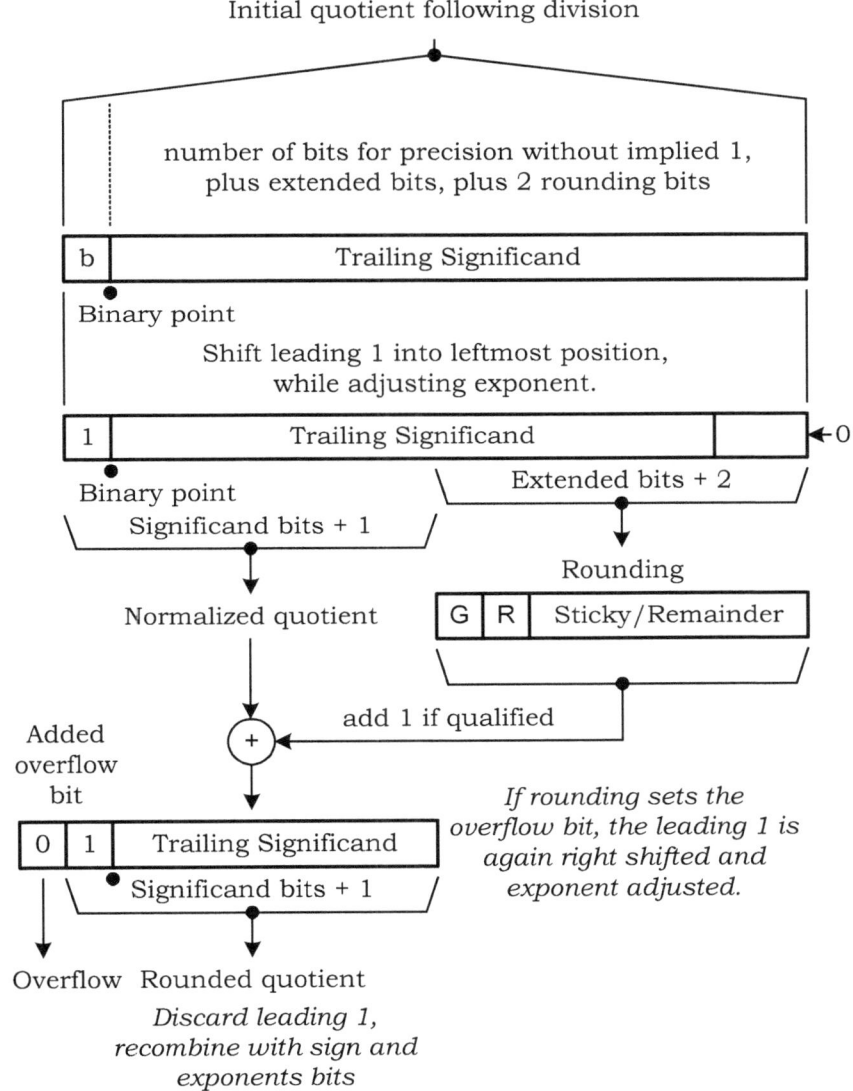

Figure 5

Figure 5 illustrates the steps for finalizing the quotient from its initial state. The initial quotient value is left shifted, while adjusting the exponent, until the most significant 1 is to the left of the binary point. The two most significant bits remaining in the lower half of the trailing significand serve as guard and round. Any remaining lower bits can be used in conjunction with the last remainder to produce the sticky bit. See section *4.4 The Sticky Bit*. Rounding is further qualified by the LSB of the normalized quotient.

If, qualified, 1 is added to the normalized quotient, and afterwards the leading 1 is again repositioned to the left of the binary-point. In the diagram an extra bit is added to the left in the event the rounding step causes an overflow. This only qualifies as an error if the exponent is at its upper bounds.

Afterwards, the rounded product is stripped of its leading '1', making it implied, then the exponent and sign are recombined with the significand into the floating-point format.

Overflow or underflow occurs when either the upper or lower bounds of the exponent (bias) are exceeded, or the significand is zero after rounding.

4.3 Rounding

Rounding the quotient following division is a complicated subject. There are many white papers detailing the advantages of certain approaches and the direction and magnitude of error incurred.

Terms like infinitely precise, the size of the resulting quotient in relationship to the size operands, the length or run of ones or zeros following the round-bit and the sign of the remainder, all contribute to error direction and magnitude. Collectively, these issues are reserved for those that specialize in the subtle effects of rounding, in particular, rounding during division, and are beyond the scope of this book.

Rounding bias is also a subject of concern by many, meaning positive and negative numbers are treated differently. The best approach is to have their effects cancel each other out (sign symmetric), or have errors accumulate in a known direction.

Additional information on common modes of rounding are provided in the "STATE MACHINE IN VHDL *Multipliers* Vol. 2" book sections 4.3 and 4.4, and are not repeated here.

All examples contained within this book employ round-to-nearest-even and GRS, except where otherwise stated. Conceptual information provided in previous sections 4.1 and 4.2 of this book, allow the designer to implement any of the fore mentioned modes.

> Warning: *in division, if the resulting quotient is an irrational number (the pattern of repeating or non-repeating bits goes on forever), the bit pattern will differ between two's complement positive and negative numbers, thus the GRS rounding strategy will have a different result depending on the sign of the quotient.*

4.4 The Sticky Bit

Generating the sticky bit is easy when using sign-magnitude representation, because the quotient and remainder are always positive. Sticky is set to '1' whenever there is a none-zero remainder. However, it is much more complicated when using signed division and two's complement representation.

The sign of the initial quotient determines whether a sticky value of '1' or '0' represents an increase in magnitude (away from zero). The sign of the remainder, on the other hand, determines how the remainder is evaluated -- for the presence of ones or for the presence of zeros.

Rescinding all other rules: if the remainder is zero, including the sign bit, sticky must be zero regardless. Secondary rules are as follows:

- If the remainder is positive, then the presence of ones means the quotient is larger in magnitude and sticky should be set. If the remainder is negative, then the presence of any zeros means the quotient is larger in magnitude and sticky should be set.

- What sticky is set to depends on the sign of the initial quotient. If the initial quotient is positive, sticky should be set to '1' otherwise '0'. If the initial quotient is negative sticky should be set to '0' otherwise '1'.

4.5 Divide By Zero

While unknown by most, what to do when faced with a zero divisor is actually not completely settled. For most in the industry, this event generates an exception or sets the quotient to a certain value as in the IEEE standard, or both.

Unless there is a good reason, industrial conventions and standards should be followed.

5 Simple Sequential Division

Basic sequential dividers are the simplest to implement, easiest to understand, and closely follow the pencil-paper approach. Its shortcomings include lengthy execution time, operating on positive numbers only, and only single quotient bit is retired at a time.

As shown in Figure 6, the divisor is subtracted from the MSB of the dividend, resulting in a remainder. The next dividend bit is then passed down and appended to the remainder, which again is subtracted by the divisor. This process is repeated through the last dividend bit.

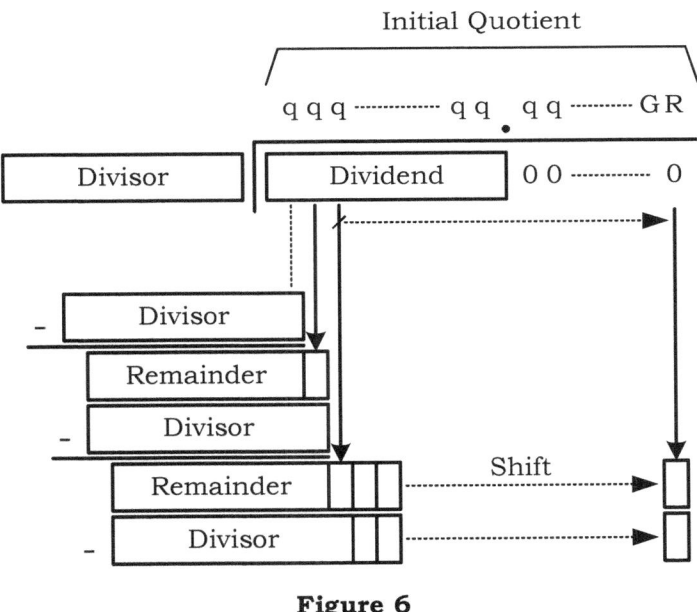

Figure 6

If the subtraction creates a borrow, meaning the remainder is less than the divisor, the corresponding quotient bit is set to a zero and the remainder is not updated, otherwise it is set to a one and the difference is retained for the next step. This closely follows the non-restoring algorithm approach. Obtaining the integer portion of the quotient requires a number of cycles equal to the number of bits in the operands. Obtaining fractional and rounding bits require more operations. Figure 7 represents the logic design of this algorithm.

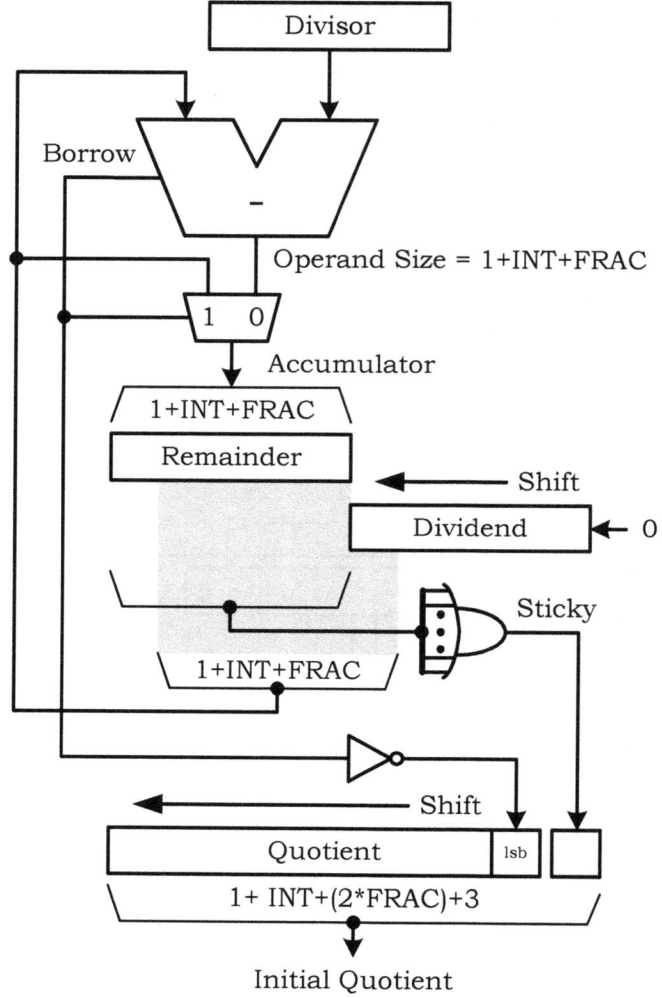

Figure 7

Note: *since this algorithm only operates on positive numbers, the sign bit need not be included. However, leaving it in simplifies the design.*

First, all but the MSB of the remainder register along with the MSB of the dividend is fed back into the subtractor, which is subtracted by the divisor. This grouping mimics a single bit left-shift. The resulting borrow signal is either set or cleared. If set, remainder register is updated with the left-shifted group and the LSB of the quotient register is cleared. If the resulting borrow is cleared, the remainder register is updated with the difference from the subtractor and the LSB of the quotient is set.

Concurrently, the dividend register is left-shifted, replacing its MSB with the next lower bit; the quotient register is left-shifted while updating its LSB with the inverted borrow signal.

The number of clock cycles required to complete the operation is equal to the number of bits in the quotient register, which is the operand size, plus the number of fractional bits in the operand, plus two bits for rounding. ORing the content of the remainder is used to set the sticky bit.

Note: *see section 4 for Normalizing, Rounding, and Bounds.*

The example state machine provided illustrates the *Simple Sequential Divider.*

- Both operands (divisor and dividend) and quotient are in the same fixed-point format.
- Format is scalable Qm.n, integer and fractional bit lengths as generic parameters, defaulting to Q31.32, with a range to Q63.64.
- Signed two's complement numbers are supported for both the operands and quotient. All conversions are handled and managed internally by the state machine.
- Rounding can be enabled or disabled, but defaults to round-to-nearest-even.

As configured for Q31.32, test builds were run using Xilinx ISE. The following performance was obtained with corresponding parts.

Xilinx Spartan XC3s500e greater than 92MHz
Xilinx Virtex XC5vlx30 greater than 233MHz

```vhdl
----------------------------------------------------
--
--  SequentialDivider.vhd
--
----------------------------------------------------
library IEEE;
use IEEE.std_logic_1164.all;
use IEEE.numeric_std.all;

entity SequentialDivider is
--
--    Qmn fixed point format is used.
--
--            sign        binary point
--             |              |
--    format <s>(integer bits).(fractional bits)
--             _____/_____/
--                 INT_SIZ         FRAC_SIZ
--
--
generic (INT_SIZ: integer range 0 to 63 := 31;
         FRAC_SIZ: integer range 0 to 63 := 32;
         ROUNDING: std_logic := '1');
port
(
    clk: in std_logic; -- system clock
    rst: in std_logic;  -- system reset (must be synchronous)
    -- inputs
    start: in std_logic; -- start division
    divisor: in signed((1 + INT_SIZ + FRAC_SIZ)-1 downto 0);
    dividend: in signed((1 + INT_SIZ + FRAC_SIZ)-1 downto 0);
    -- ouputs
    cmplt: out std_logic; -- division complete
    ovrflw: out std_logic; -- overflow error
    udrflow: out std_logic; -- overflow error
    quotient: out signed((1 + INT_SIZ + FRAC_SIZ)-1 downto 0)
);
end SequentialDivider;

architecture RTL of SequentialDivider is

------------------------
-- Declared constants
------------------------
```

```vhdl
constant D_SIZ: integer := 1 + INT_SIZ + FRAC_SIZ; -- data word size
constant R_SIZ: integer := 1 + INT_SIZ + FRAC_SIZ; -- remainder size
constant Q_SIZ: integer := 1 + INT_SIZ + (2*FRAC_SIZ) + 2; -- quotient size

constant D_MSB: integer := D_SIZ-1; -- data word msb bit position
constant R_MSB: integer := R_SIZ-1; -- remainder msb bit position

constant D_OVR: integer := D_SIZ; -- data word overflow bit position

constant Q_LSB: integer := 2; -- quotient lsb bit position
constant Q_MSB: integer := D_MSB+Q_LSB; -- quotient msb bit position

----------------------
-- Declared signals
----------------------
signal w1: signed(D_MSB downto 0) := (others=>'0');
signal w2: signed(D_MSB downto 0) := (others=>'0');
signal remainder: unsigned(R_MSB downto 0) := (others=>'0');
signal grs: unsigned(2 downto 0) := (others=>'0');
signal quo: signed(Q_SIZ-1 downto 0) := (others=>'0');

signal a: unsigned(D_MSB downto 0) := (others=>'0');
signal b: unsigned(D_MSB downto 0) := (others=>'0');
signal dout: unsigned(D_MSB downto 0) := (others=>'0');
signal bout: unsigned(D_MSB downto 0) := (others=>'0');
signal bin: unsigned(D_MSB downto 0) := (others=>'0');

signal cnt: integer range 0 to Q_SIZ-1 := 0;
signal sign: std_logic := '0';
signal busy: std_logic := '0';


----------------------
-- Enumeration lists
----------------------
type sm_def is
(
    RESET,
    START_DIV,
    DIV,
    DIV2,
    DIV3,
    ROUND,
    ROUND2
);
```

```vhdl
signal state: sm_def := RESET;

------------------------------------ module code ----------------------------
begin


----------------------------------------
--
--   Sequential Divider state machine
--
----------------------------------------
process(rst,clk)
begin

    if(rst='1') then

        -- working registers
        w1  <= (others=>'0');
        w2  <= (others=>'0');
        remainder  <= (others=>'0');
        grs <= (others=>'0');
        quo <= (others=>'0');

        -- local signals
        sign <= '0';
        cnt <= 0;

        -- handshake signals
        busy <= '0';
        ovrflw <= '0';
        udrflw <= '0';

        -- states
        state <= RESET;

    elsif rising_edge(clk) then
        --
        --   state machine body
        --
        case state is
            -- reset state
            when RESET =>
                state <= START_DIV;
            --
            -- divider body
```

```vhdl
--
when START_DIV =>
    if(start = '1') then
        w1 <= divisor;
        w2 <= dividend;
        busy <= '1';
        ovrflw <= '0';
        udrflw <= '0';
        cnt <= 0;
        state <= DIV;
    end if;
-- division operation (w1->divisor, w2->dividend)
when DIV =>
    -- clear remainder
    remainder <= (others=>'0');
    grs <= (others=>'0');
    -- save resultant sign
    sign <= w1(w1'high) xor w2(w2'high);
    -- negate w1 operand
    if(w1(w1'high) = '1') then
        w1 <= (not w1) + 1;
    end if;
    -- negate w2 operand
    if(w2(w2'high) = '1') then
        w2 <= (not w2) + 1;
    end if;
    state <= DIV2;
when DIV2 =>
    -- left-shift dividend for next cycle
    w2 <= w2(w2'high-1 downto 0)&'0';
    -- keep difference if there is no borrow, otherwise ignore
    if(bout(bout'high) = '0') then
        remainder <= dout; -- difference out from subtractor
    else
        remainder <= a; -- remainder left-shifted
    end if;
    -- set quotient bit and left-shift
    quo <= quo(quo'high-1 downto 0)&(not bout(bout'high));
    -- sequence counter
    if(cnt < Q_SIZ-1) then
        cnt <= cnt + 1;
    else
        state <= DIV3;
    end if;
```

```vhdl
            when  DIV3 =>
                -- normalize quotient back into quotient register
                quo(D_OVR downto 0) <= '0'&quo(Q_MSB downto Q_LSB);
                -- save guard and round bits
                grs(2 downto 1) <= unsigned(quo(1 downto 0));
                -- set sticky if remainder is not zero
                if(remainder /= 0) then
                    grs(0) <= '1';
                end if;
                -- set overflow flag
                if(quo(quo'high downto Q_MSB) /= 0) then
                    ovrflw <= '1';
                    busy <= '0';
                    state <= START_DIV;
                else
                    state <= ROUND;
                end if;
            -- rounding for division
            when ROUND =>
                -- round result to nearest even number
                if (ROUNDING = '1' and(grs > 4 or (grs = 4 and quo(0) = '1'))) then
                    quo(D_OVR downto 0) <= '0'&quo(D_MSB downto 0) + 1;
                end if;
                state <= ROUND2;
            when ROUND2 =>
                if(quo(D_OVR downto D_MSB) /= 0) then
                    ovrflw <= '1';
                elsif(quo(D_MSB-1 downto 0) = 0 ) then
                    udrflw <= '1';
                elsif(sign = '1') then
                    quo(D_MSB downto 0) <= (not quo(D_MSB downto 0)) + 1;
                end if;
                busy <= '0';
                state <= START_DIV;
            when others =>
                state <= RESET;
        end case;
    end if;

end process;

-----------------------
--
```

```vhdl
-- Subtractor circuit
--
-----------------------
-- remainder+msb of dividend
a <= remainder(remainder'high-1 downto 0)&w2(w2'high);
b <= unsigned(w1); -- divisor

-- full subtractors
gen_sub: for i in 0 to D_MSB generate
begin
    -- connect borrows
    gen_bin0:if(i = 0) generate
        bin(i) <= '0';
    end generate;
    gen_bin:if(i > 0) generate
        bin(i) <= bout(i-1);
    end generate;
    -- compute difference out
    dout(i) <= (a(i) xor b(i)) xor bin(i);
    -- compute borrow out
    bout(i) <= ((not a(i)) and bin(i))or ((not a(i)) and b(i)) or (b(i) and bin(i));

end generate;

-- output signals
cmplt <= (not start) and (not busy);
quotient <= quo(D_MSB downto 0);

end RTL;
```

6 Signed Sequential Division

It has been said: "there is no easy way to do signed division", and that is certainly the case. Division with positive operands is easier, unless reducing the execution time by a few clocks is necessary.

The up-side of signed division is not having to convert negative operands to positive ones before execution, and possibly convert the resulting positive quotient to a negative one. The down-side is a reduction in obtainable clock speed due to the extra logic driven by the carry chain, gating to support conversion cycles, and the increased complexity of the design itself, especially in obtaining accurate rounding.

Two design strategies are covered in this chapter. The first relies on a more conventional approach known as *Signed Sequential Using Conventional Wisdom*. The second and more academic design, is referred to as *Signed Sequential Using On-The-Fly-Conversion*.

Both methods are non-restoring. Both are sequential and retire a single quotient bit at a time. Both require no external sign conversion. Both require a correction step at the end. Although very similar, one employs a shift-subtract algorithm, and the other a shift-add/shift-subtract algorithm.

Signed Sequential Using On-The-Fly-Conversion is slightly faster because it can achieve a higher clock rate, but *Signed Sequential Using Conventional Wisdom* can do the same operation in one less clock cycle. The speed results can depend greatly on the targeted device (capabilities within the device) due to the structural differences between the two designs.

Two of the greatest limiting factors are the generation of the sticky bit through the remainder correction path and the sheer size of the quotient correction adder. Both designs are represented with an intention of efficiency, but also in a way that is understandable by the reader. Focused effort in these areas can bring a fair measure of improvement in speed and resource use.

6.1 Signed Sequential Using Conventional Wisdom

Conventional wisdom merely means that all sub-operations within the algorithm are done with two's complement math, not with more specialized techniques.

Operands in two's complement representation are used as is with no pre-conversion required. However, post-conversion is required for the quotient, but is embedded in the algorithm and completes following the retirement of the last quotient bit.

The table below shows how the signs of the operands and quotient play a role in how the algorithm operates.

Negate Divisor	Operand Signs		Quotient Sign		Quotient Bit	Correct ?
	Dividend	Divisor	Resulting	Required		
Y	+	+	+	+	INV Borrow	No
N	+	−	+	−	Borrow	Yes (1)
N	−	+	−	−	INV Borrow	Yes (2)
Y	−	−	−	+	Borrow	Yes (1,2)

Note 1: *The quotient must be corrected by performing a two's complement negation. The first phase is done as bits are retired by using a borrow instead of the inverted borrow. The second phase of the negation, adding 1, is done at the end, but only if a sum of zero was detected at anytime during the operation.*
Note 2: *The remainder must be corrected by adding the value of the divisor back, but only if a sum of zero was detected during the operation.*

Basic sequential division involves repetitive subtractions from the current partial remainder by the divisor. If a borrow occurs, the quotient bit is set false and the current remainder is offset with the next dividend bit. If a borrow does not occur, the quotient bit is set true and the difference from the subtraction becomes the new partial remainder.

These steps are easy and straightforward using two's complement arithmetic. The divisor is always negated to be the opposite sign of the dividend, which is then added to each partial remainder to perform the subtraction. The problem is that the resulting quotient always has the sign of the dividend and must be negated in two cases to maintain the like-sign-positive and unlike-sign-negative convention, as shown in gray in the table above.

The solution is whenever the divisor is negative, the resulting quotient must be negated at the end of the operation. This is achieved in two steps. First, as quotient bits are retired, each quotient bit is set to the value of the borrow, not the inverted borrow, thus performing the invert phase of a two's complement negation. Then, after retiring the final quotient bit, a 1 is added to the entire quotient, completing the second phase of negation. Figures 8 and 9 represent such an implementation.

Note: *The grayed components in Figures 8 and 9 represent the remainder correction and sticky decode circuits, which are covered at the end of this section.*

Rules governing the basic algorithm:

1). All bits in the remainder are initialized to the sign of the dividend.

2). The MSB or sign bit of the sum (output of adder), which is the result of the current subtraction, serves as the borrow flag, which its polarity is determined by the sign of the dividend.

If the dividend is (+) then borrow is true if sum(MSB) = '1'

If the dividend is (-) thenborrow is true if sum(MSB) = '0'

3). The borrow bit is used to set each quotient bit.

If the divisor is (+) then the inverted borrow becomes the quotient bit.

If the divisor is (-) then the borrow becomes the quotient bit, phase 1 in the quotient negation.

4). The cycle following the retirement of the final quotient bit is used to correct the final quotient and remainder and generate the sticky bit.

Steps of algorithm:

Input 'b' to the adder is the divisor, which is negated whenever the signs of the dividend and divisor match to perform the two's complement subtraction by adding.

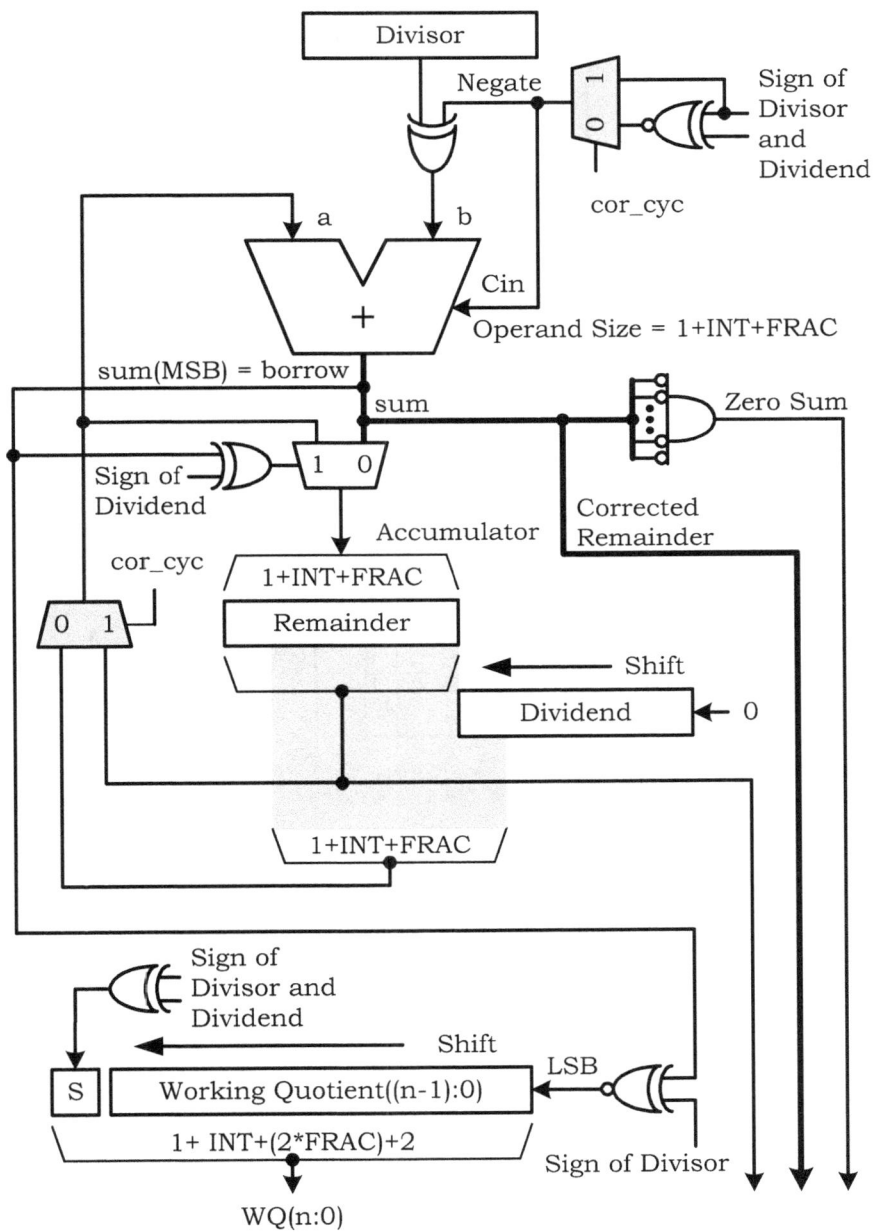

Figure 8

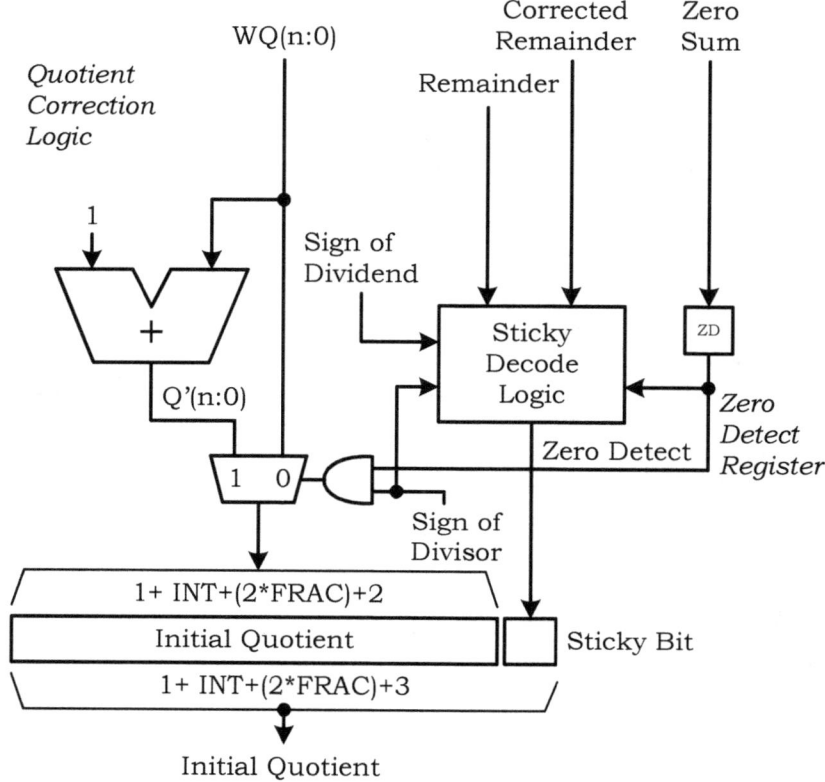

Figure 9

If the resulting add produces a borrow status, the remainder is updated with input 'a' to the adder, which is the current remainder offset with the next dividend bit. If a borrow status is false, then the remainder register is updated with the sum of the adder, which is the difference between inputs 'a' and 'b'.

During each cycle, the LSB of the working quotient register is updated, based on the combination (XNOR) borrow and the divisor sign, while left-shifting the entire register.

As each cycle occurs, the remainder register (excluding its MSB) along with the MSB of the working dividend register, are fed back into input 'a' of the adder. At the same time the dividend register is left-shifted, putting the next bit into position for the next cycle.

Figure 9 represents the final stage of the divider, which carries out the second step in the quotient negation, if needed, following the operation. During this time, the value of the sticky bit is also determined – the remainder may require correction.

> Note: *While an extra clock cycle is needed to perform correction, the overall number of clock cycle was reduced by 1 since the MSB of the quotient is known immediately (dividend sign xor divisor sign).*

The signal referred to in Figure 8 as *cor_cyc* (*Correction Cycle*) gates the final remainder back through the adder, as well as, properly adding the divisor back to its value, whether it is used or not.

Rules governing correction:

1). The working quotient is either passed through as-is or is corrected by adding 1, which is the second phase in the two's complement negation. This occurs only if the divisor is negative and a zero sum was detected at any time during the operation.

2). The final remainder, which may need to be corrected as well, is used to determine the value of the sticky bit. The remainder only needs to be corrected if the sign of the dividend is negative and a zero sum was detected during the operation.

3). Detection of a zero sum determines whether the remainder or the corrected remainder are used in the decode of the sticky bit.

4). Section *4.4 The Sticky Bit* explains the rules in determining its value. The signs of the dividend and divisor are used to qualify what the quotient sign should be. The sign of the remainder will track with the sign of the dividend after correction unless the correction reduces the remainder to a value zero.

The number of clock cycles required to complete the operation is equal to the number of bits in the quotient register, which is the operand size, plus the number of fractional bits in the operand, plus several bits (2) for rounding.

Optional Improvements

1). It is obvious that remainder correction and the generation of the sticky bit consumes resources and impacts performance.

- If greater accuracy in rounding is not needed, the sticky bit can be generated in the same way as the guard and round bits are, or all three bits, GRS need not be generated at all.
- Restructuring the sticky bit decode logic for more optimal synthesis can be done as well.

2). The corrected remainder value at the output of the adder can be registered, which will reduce the path to the sticky bit. Alternatively, simply register the corrected value in the remainder while maintaining a copy, using that path for correction .Either approach will require an extra clock cycle, but will increase the clock speed.

3). If the path to the sticky bit has been improved, the next improvement with the greatest return involves the quotient correction circuits. While only comprised of a simple adder and mux, the width of the adder can be huge because it includes the entire initial quotient. When normalized, the upper bits are only used to detect overflow. The upper bits need not be included in the carry chain. The overflow logic within the correction adder can be included.

Example Design

The example state machine provided illustrates the *Signed Sequential Divider Using Conventional Wisdom.*

- Both operands (divisor and dividend) and quotient are in the same fixed-point format.
- The format is scalable Qm.n, integer and fractional bit lengths as generic parameters, defaulting to Q31.32, with a range to Q63.64.
- Signed two's complement numbers are supported for both the operands and quotient.
- Rounding can be enabled or disabled, but defaults to round-to-nearest-even.

As configured for Q31.32, test builds were run using Xilinx ISE. The following performance figures were obtained with corresponding parts.

    Xilinx Spartan XC3s500e greater than 83MHz
    Xilinx Virtex XC5vlx30 greater than 182MHz

```vhdl
--------------------------------------------------------------------------
--
--    SignedSequentialDivider_cw.vhd
--
--    Signed sequential divider using conventional wisdom (cw).
--
--------------------------------------------------------------------------
library IEEE;
use IEEE.std_logic_1164.all;
use IEEE.numeric_std.all;

entity SignedSequentialDivider is
--
--    Qmn fixed point format is used.
--
--          sign       binary point
--           |              |
--    format <s>(integer bits).(fractional bits)
--           _____/_____/
--              INT_SIZ         FRAC_SIZ
--
--
generic (INT_SIZ: integer range 0 to 63 := 31;
         FRAC_SIZ: integer range 0 to 63 := 32;
         ROUNDING: std_logic := '1');
port
(
    clk: in std_logic; -- system clock
    rst: in std_logic;  -- system reset (must be synchronous)
    -- inputs
    start: in std_logic; -- start division
    divisor: in signed((1 + INT_SIZ + FRAC_SIZ)-1 downto 0);
    dividend: in signed((1 + INT_SIZ + FRAC_SIZ)-1 downto 0);
    -- ouputs
    cmplt: out std_logic; -- division complete
    ovrflw: out std_logic; -- overflow error
    udrflw: out std_logic; -- underflow error
    quotient: out signed((1 + INT_SIZ + FRAC_SIZ)-1 downto 0)
);
end SignedSequentialDivider;

architecture RTL of SignedSequentialDivider is
```

```
--------------
--  Functions
--------------
function OnesPresent(num: unsigned) return std_logic is
variable ret: std_logic := '0';
begin

    for i in num'high downto num'low loop
        if(num(i) = '1') then
            ret := '1';
            exit;
        end if;
    end loop;
    return(ret);

end function;

function ZerosPresent(num: unsigned) return std_logic is
variable ret: std_logic := '0';
begin

    for i in num'high downto num'low loop
        if(num(i) = '0') then
            ret := '1';
            exit;
        end if;
    end loop;
    return(ret);

end function;


------------------------
--  Declared constants
------------------------
constant D_SIZ: integer := 1 + INT_SIZ + FRAC_SIZ; -- data word size
constant R_SIZ: integer := 1 + INT_SIZ + FRAC_SIZ; -- remainder size
constant Q_SIZ: integer := 1 + INT_SIZ + (2*FRAC_SIZ)+ 2; -- quotient size

constant D_MSB: integer := D_SIZ-1; -- data word msb bit position
constant R_MSB: integer := R_SIZ-1; -- remainder msb bit position

constant D_OVR: integer := D_SIZ; -- data word overflow bit position

constant Q_LSB: integer := 2; -- quotient lsb bit position
```

constant Q_MSB: integer := D_MSB+Q_LSB; -- quotient msb bit position

----
-- Declared signals
----
signal w1: signed(D_MSB downto 0) := (others=>'0');
signal w2: signed(D_MSB-1 downto 0) := (others=>'0');

signal remainder: unsigned(R_MSB downto 0) := (others=>'0');
signal grs: unsigned(2 downto 0) := (others=>'0');

signal work_quo: signed(Q_SIZ-1 downto 0) := (others=>'0');
signal q_prime: signed(Q_SIZ-1 downto 0) := (others=>'0');
signal quo: signed(Q_SIZ-1 downto 0) := (others=>'0');

signal a: unsigned(D_MSB downto 0) := (others=>'0');
signal b: unsigned(D_MSB downto 0) := (others=>'0');
signal sum: unsigned(D_MSB downto 0) := (others=>'0');

signal cnt: integer range 0 to Q_SIZ-2 := 0;
signal negate: std_logic := '0';
signal dividend_sign: std_logic := '0';
signal divisor_sign: std_logic := '0';
signal d_xnor_d: std_logic := '0';
signal borrow: std_logic := '0';
signal busy: std_logic := '0';
signal cor_cyc: std_logic := '0';
signal zero_det: std_logic := '0';

signal zero_sum: std_logic := '0';
signal presence_of_ones: std_logic := '0';
signal presence_of_zeros: std_logic := '0';
signal cor_presence_of_zeros: std_logic := '0';
signal sticky: std_logic := '0';

alias cor_remainder is sum;

```vhdl
------------------------
-- Enumeration lists
------------------------
type sm_def is
(
    RESET,
    START_DIV,
    DIV,
    DIV2,
    DIV3,
    ROUND,
    ROUND2
);
signal state: sm_def := RESET;

------------------------------------ module code ----------------------------
begin
-------------------------------------------
--
--   Signed Sequential Divider state machine
--
-------------------------------------------
process(rst,clk)
begin

    if(rst='1') then

        -- working registers
        w1  <= (others=>'0');
        w2  <= (others=>'0');
        remainder  <= (others=>'0');
        work_quo <= (others=>'0');
        quo <= (others=>'0');
        grs <= (others=>'0');

        -- local signals
        dividend_sign <= '0';
        divisor_sign <= '0';
        d_xnor_d <= '0';
        cnt <= 0;
        cor_cyc <= '0';
        zero_det <= '0';

        -- handshake signals
```

```vhdl
            busy <= '0';
            ovrflw <= '0';
            udrflw <= '0';

            -- states
            state <= RESET;

        elsif rising_edge(clk) then

            -- one clock signals
            cor_cyc <= '0';

            --
            --   state machine body
            --
            case state is
                -- reset state
                when RESET =>
                    state <= START_DIV;
                --
                --  divider body
                --
                when START_DIV =>
                    if(start = '1') then
                        w1 <= divisor;
                        divisor_sign <= divisor(divisor'high);
                        w2 <= dividend(dividend'high-1 downto 0);
                        dividend_sign <= dividend(dividend'high);
                        d_xnor_d <= divisor(divisor'high) xnor dividend(dividend'high);
                        -- prime remainder register with dividend sign
                        for i in remainder'high downto 0 loop
                            remainder(i) <= dividend(dividend'high);
                        end loop;
                        -- initialize variables
                        work_quo <= (others=>'0');
                        quo <= (others=>'0');
                        grs <= (others=>'0');
                        -- other
                        zero_det <= '0';
                        busy <= '1';
                        ovrflw <= '0';
                        udrflw <= '0';
                        cnt <= 0;
                        state <= DIV;
```

```vhdl
        end if;
-- division sequence

when DIV =>
    -- left-shift working dividend
    w2 <= w2(w2'high-1 downto 0)&'0';
    -- update remainder based on original dividend sign and current
    -- borrow
    if(dividend_sign = borrow) then
        -- keep results as new remainder
        remainder <= sum;
    else
        -- do not keep result, feedback remainder left-shifted
        -- with new msb dividend bit
        remainder <= a;
    end if;
    -- set working quotient lsb bit while left-shifting all other bits
    work_quo <= work_quo(work_quo'high-1 downto work_quo'low)
                &(divisor_sign xnor borrow);
    -- override msb
    work_quo(work_quo'high) <= (divisor_sign xor dividend_sign);
    -- detect a zero output from the primary adder
    zero_det <= zero_det or zero_sum;
    -- sequence counter
    if(cnt < Q_SIZ-2) then
        cnt <= cnt + 1;
    else
        -- completed, start correction cycle
        cor_cyc <= '1';
        state <= DIV2;
    end if;
-- correction cycle
when DIV2 =>
    -- correct quotient if divisor is negative
    if(divisor_sign = '1' and zero_det = '1') then
        quo <= q_prime; -- corrected quotient
    else
        quo <= work_quo; -- as is
    end if;
    -- corrected sticky bit
    grs(0) <= sticky;
    state <= DIV3;

when DIV3 =>
```

```vhdl
                -- normalize quotient back into quotient register
                quo(D_OVR downto 0) <= quo(Q_MSB)&
                                            quo(Q_MSB downto Q_LSB);
                state <= ROUND;
                -- set overflow flag based in sign
                for i in quo'high downto Q_MSB loop
                    if(quo(i) /= quo(Q_MSB)) then
                        ovrflw <= '1';
                        busy <= '0';
                        state <= START_DIV;
                    end if;
                end loop;
                -- save guard and round bits from retired quotient bits
                grs(2 downto 1) <= unsigned(quo(1 downto 0));
            -- rounding for division
            when ROUND =>
                -- round result to nearest even number
                if (ROUNDING = '1') then
                    if(grs > 4 or (grs = 4 and quo(0) = '1')) then
                        -- round up
                        quo(D_OVR downto 0) <= quo(D_OVR downto 0) + 1;
                    else
                        null;-- truncate
                    end if;
                end if;
                state <= ROUND2;
            when ROUND2 =>
                -- check for overflow from rounding
                if(quo(D_OVR) /= quo(D_MSB)) then
                    ovrflw <= '1';
                -- data portion of operand equal to zero qualifies as an underflow
                elsif(quo(D_MSB downto 0) = 0) then
                    udrflw <= '1';
                end if;
                busy <= '0';
                state <= START_DIV;
            when others =>
                state <= RESET;
        end case;
    end if;

end process;
```

-- Adder circuits

-- remainder with next msb of dividend or remainder as is during correction cycle
a <= remainder(remainder'high-1 downto 0)&w2(w2'high) when cor_cyc = '0'
       else remainder;

-- subtract or add divisor
negate <= d_xnor_d when cor_cyc = '0' else divisor_sign;
b <= unsigned((not(w1))+1) when negate = '1' else unsigned(w1);

-- sum and borrow outputs of adder
sum <= a + b;
borrow <= sum(sum'high);

-- Quotient correction adders
----
q_prime <= work_quo(work_quo'high downto work_quo'low) + 1;

-- Sticky decode logic
----
zero_sum <= '1' when sum = 0 else '0';

-- check for ones and zeros without the sign bit
presence_of_ones <= OnesPresent(remainder(remainder'high-1 downto 0));
presence_of_zeros <= ZerosPresent(remainder(remainder'high-1 downto 0));
cor_presence_of_zeros <= ZerosPresent(cor_remainder(remainder'high-1 downto 0));

-- sticky decode logic
stky_gen: process(dividend_sign,divisor_sign,zero_det,remainder,cor_remainder,
            presence_of_ones,presence_of_zeros,cor_presence_of_zeros)
begin

       --
       -- (+) dividend
       --
       if(dividend_sign = '0') then
           -- check for zero remainder
           if(remainder = 0) then
               sticky <= '0';
           -- (+) divisor

```vhdl
    elsif(divisor_sign = '0') then
        -- 1s present then sticky is 1 for positive quotient
        sticky <= presence_of_ones;
    -- (-) divisor
    else
        -- 1s present then sticky is 0 for negative quotient
        sticky <= not presence_of_ones;
    end if;
--
-- (-) dividend  (correction my be required)
--
else
    -- (+) divisor
    if(divisor_sign = '0') then
        -- correction required
        if(zero_det = '1') then
            -- corrected 0s present then sticky is 0 for negative quotient
            if(cor_remainder = 0) then
                sticky <= '0';
            else
                sticky <= not cor_presence_of_zeros;
            end if;
        -- no correction required
        else
            -- 0s present then sticky is 0 for negative quotient
            if(remainder = 0) then
                sticky <= '0';
            else
                sticky <= not presence_of_zeros;
            end if;
        end if;
    -- (-) divisor
    else
        -- correction required
        if(zero_det = '1') then
            -- corrected 0s present then sticky is 1 for positive quotient
            if(cor_remainder = 0) then
                sticky <= '0';
            else
                sticky <= cor_presence_of_zeros;
            end if;
        -- no correction required
        else
            -- 0s present then sticky is 1 for positive quotient
```

```vhdl
                    if(remainder = 0) then
                        sticky <= '0';
                    else
                        sticky <= presence_of_zeros;
                    end if;
                end if;
            end if;
        end if;

end process;

-- output signals
cmplt <= (not start) and (not busy);
quotient <= quo(D_MSB downto 0);

end RTL;
```

6.2 Signed Sequential Using On-The-Fly-Conversion

As with the *Conventional Wisdom* algorithm in the previous section, no pre-conversion of the two's complement operands is required. Unlike the previous algorithm, post conversion utilizes a technique called *on-the-fly-conversion* which converts a number from one numeric representation to another as each quotient bit is retired. In this case a conversion from signed digit-set {-1, +1} to two's complement representation.

> Note: *The signed digit representation used in this section is not a true signed digit, rather pseudo. Later chapters employ them in their proper form.*

Sub-operations in the divide algorithm are done using two's complement math which generate an encoded Signed-Digit (SD) for each bit, which is then converted back to a two's complement result.

Signs of both the current partial remainder and the divisor determine the sub-operation for the current cycle, as shown in the table below.

Signs		Negate	Quotient	New
Remainder	Divisor	Divisor	SD Bit	Remainder
+	+	Y	+1	Remainder - Divisor
+	-	N	-1	Remainder + Divisor
-	+	N	-1	Remainder + Divisor
-	-	Y	+1	Remainder - Divisor

Note: *The final quotient will require correction if either an intermediate zero occurred, all cycles except the last one, or if the final remainder and dividend signs do not match.*

The algorithm's sequence is simple. The sign bits of the remainder and divisor are evaluated. If they are the same, then the quotient bit is set to {+1} and the divisor is subtracted from the remainder. If they are not the same, then the quotient bit is set to {-1} and the divisor is added to the remainder. In either case, the difference or the sum becomes the new partial remainder.

Figure 10 represents on-the-fly-conversion. This technique is normally implemented to operate as bits are retired. However, because a correction cycle is needed to support the divide algorithm, it occurs during that cycle and is used more as a SD to two's complement converter.

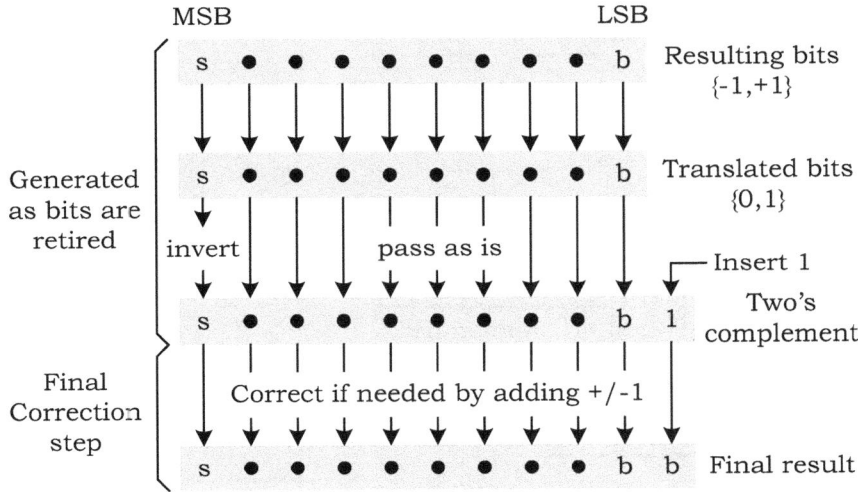

Figure 10

First, based on the sub-operations in the previous table, each bit is set to a SD {-1} or {+1}. Since only two states exist, they translate into {0} or {1}, respectively. Next, the MSB is inverted, leaving all others unchanged, then a '1' is appended as the LSB.

Lastly, if a correction step is needed, a 1 is either added or subtracted, ending in the final result. The remainder also has to be corrected to generate the proper sticky bit value.

> Note: *the final size of the initial quotient will be 1-bit larger. The extra bit serves as an overflow detection. This is not really needed since the initial quotient is normalized into the source operand sizes.*

Figures 11 and 12 represent the entire implementation.

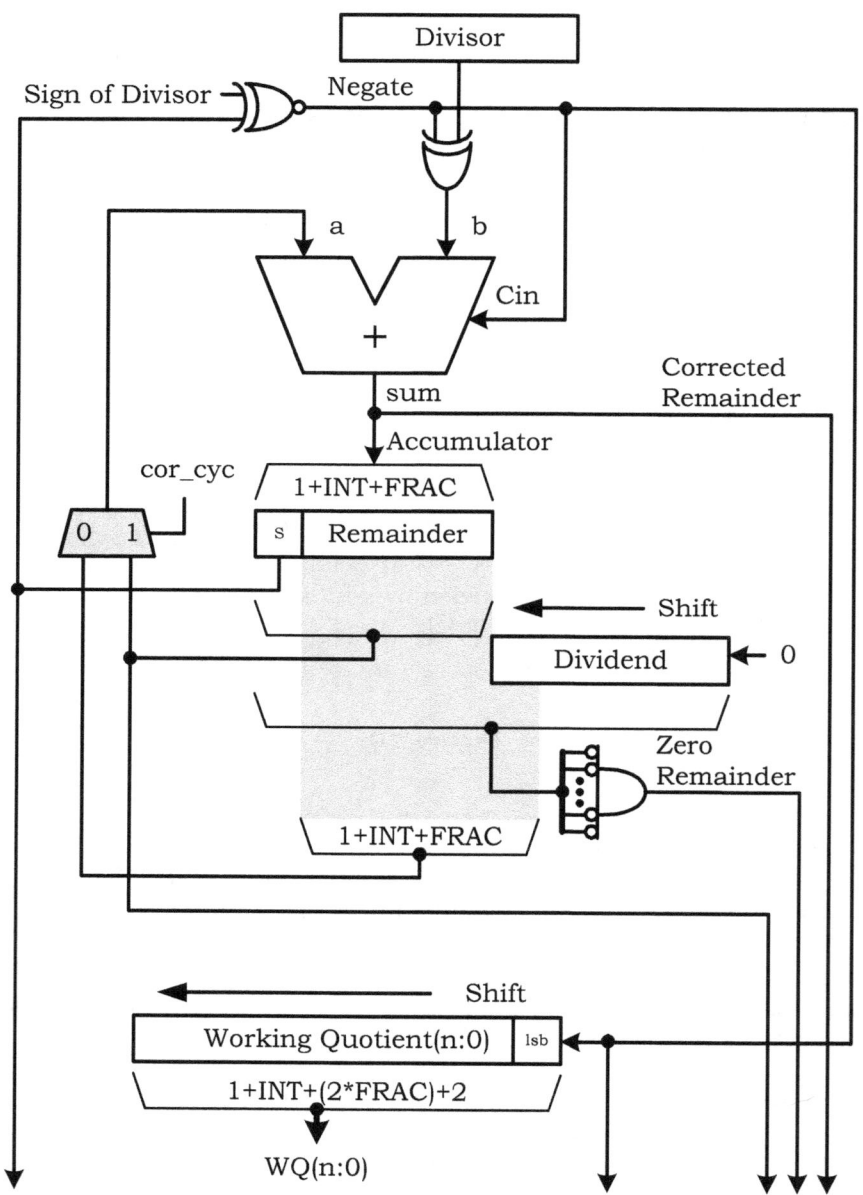

Figure 11

Note: *The grayed components in Figures 11 and 12 represent the remainder correction circuits.*

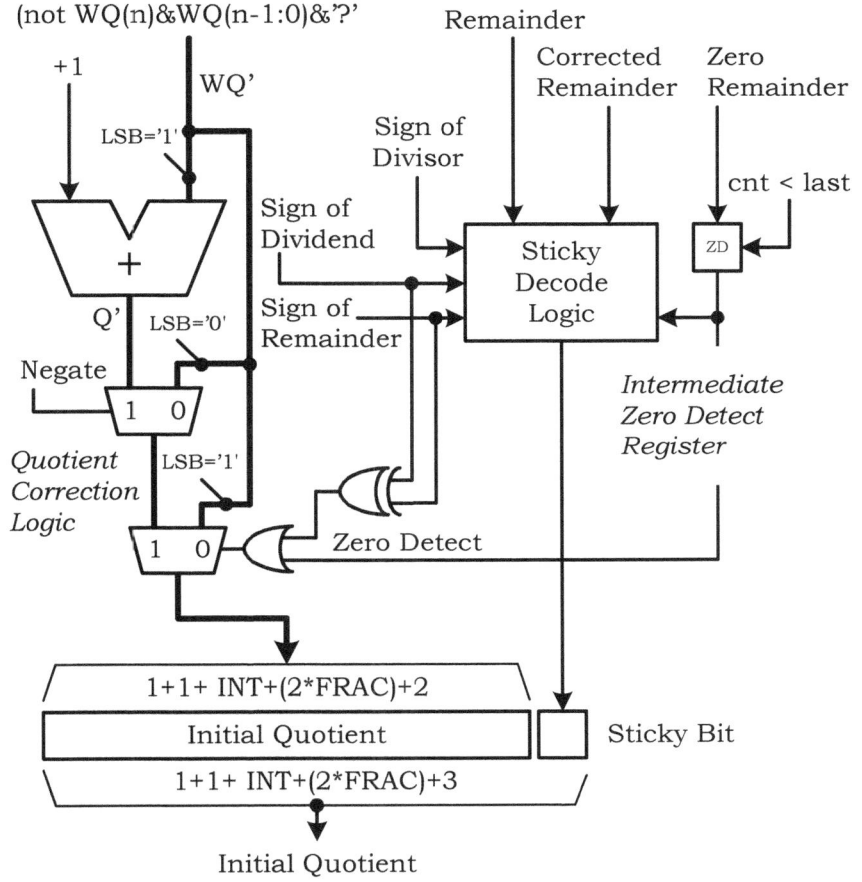

Figure 12

Rules governing the basic algorithm:

1). All bits in the remainder are initialized to the sign of the dividend.

2). The remainder and divisor signs are evaluated at the beginning of each cycle.

   If the remainder sign and divisor signs are equal

   Quotient bit = {+1}
   Remainder = Remainder – Divisor (subtract)

If the remainder sign and divisor signs are not equal

> Quotient bit = {-1}
> Remainder = Remainder + Divisor (add)

3). The cycle following the retirement of the last quotient is used to correct the quotient and remainder, and generate the sticky bit.

Steps of Algorithm:

Input 'b' to the adder is the divisor. This is negated whenever the signs of the partial remainder and divisor are equal to perform the two's complement subtraction by adding. If their signs are different then the divisor is added instead. The output of the adder becomes the new partial remainder, whether it is the sum or the difference.

Input 'a' is sourced from the current partial remainder – without the MSB and offset by the next dividend bit, while the entire working dividend register is left-shifted.

The *negate* signal is used directly to set the next quotient bit to either {-1} or {+1}, while the working quotient register is left-shifted.

Both the entire partial remainder register along with the working dividend register are monitored for a zero. If a zero occurs anytime before the last cycle, it is flagged as an intermediate zero. This is one of the two qualifications for a correction step.

The signal referred to in Figure 11 as *cor_cyc* (Correction Cycle) gates the final remainder back through the adder while adding the divisor back to its value, whether it is used or not.

Rules governing correction:

If an intermediate zero was detected during the operation or if the signs of the partial remainder and the dividend do not match, both the remainder and the working quotient must be corrected. The sticky bit is generated at the same time.

1). Correcting the working quotient is done during the clock cycle following the last quotient bit, which includes *on-the-fly-conversion*, and involves inverting the MSB and appending a '1' to the LSB.

This result is referred to as *WQ'*. Muxes selects between results. If the quotient needed correcting, the remainder will as well.

   If the signs of the final remainder and divisor are equal

      Initial quotient = WQ'&'1' + 1   (Q = Q+1)
      Corrected remainder = Final remainder - divisor.

   If the signs of the final remainder and divisor are not equal

      Initial quotient = WQ'&'0'   (Q = Q-1)
      Corrected remainder = Final remainder + divisor

   If no correction occurred

      Initial quotient = WQ'&'1' (no adding or subtracting)
      Remainder = Final remainder.

2). If correction is required, the corrected remainder is used to decode the sticky bit, otherwise the final remainder is used. Following correction, the sign of the remainder should match the sign of the dividend, or the entire remainder will be zero.

3). Decoding the sticky bit is based primarily on the signs of the dividend and divisor, which determine the final sign of the quotient and remainder. See section *4.4 The Sticky Bit* for the rules for computing the sticky bit.

The number of clock cycles required to complete the operation is equal to the number of bits in the quotient register, which is the operand size, plus the number of fractional bits in the operand, plus several bits (2) for rounding, plus 1 cycle for the combined correction and *on-the-fly-conversion*.

Optional Improvements

Use the same suggestions for the *Signed Sequential Using Conventional Wisdom* implementation in the previous section.

Example Design

The example state machine provided illustrates the *Signed Sequential Divider*.

- Both operands (divisor and dividend) and quotient are in the same fixed-point format.
- The format is scalable Qm.n, integer and fractional bit lengths as generic parameters, defaulting to Q31.32, with a range to Q63.64.
- Signed two's complement numbers are supported for both the operands and quotient.
- Rounding can be enabled or disabled, but defaults to round-to-nearest-even is used.

As configured for Q31.32, test builds were run using Xilinx ISE. The following performance figures were obtained with corresponding parts.

> Xilinx Spartan XC3s500e greater than 77MHz
> Xilinx Virtex XC5vlx30 greater than 200MHz

```vhdl
----------------------------------------------------------------------------
--
--    SignedSequentialDivider_otf.vhd
--
--    Signed sequential divider using on-the-fly(otf).
--
----------------------------------------------------------------------------
library IEEE;
use IEEE.std_logic_1164.all;
use IEEE.numeric_std.all;

entity SignedSequentialDivider is
--
--    Qmn fixed point format is used.
--
--          sign        binary point
--           |              |
--    format <s>(integer bits).(fractional bits)
--           _____/ _____/
--                INT_SIZ        FRAC_SIZ
--
--
generic (INT_SIZ: integer range 0 to 63 := 31;
         FRAC_SIZ: integer range 0 to 63 := 32;
         ROUNDING: std_logic := '1');
port
(
    clk: in std_logic; -- system clock
    rst: in std_logic;  -- system reset (must be synchronous)
    -- inputs
    start: in std_logic; -- start division
    divisor: in signed((1 + INT_SIZ + FRAC_SIZ)-1 downto 0);
    dividend: in signed((1 + INT_SIZ + FRAC_SIZ)-1 downto 0);
    -- ouputs
    cmplt: out std_logic; -- division complete
    ovrflw: out std_logic; -- overflow error
    udrflw: out std_logic; -- underflow error
    quotient: out signed((1 + INT_SIZ + FRAC_SIZ)-1 downto 0)
);
end SignedSequentialDivider;

architecture RTL of SignedSequentialDivider is
```

```vhdl
--  ---------------
--  Functions
--  ---------------
function OnesPresent(num: unsigned) return std_logic is
variable ret: std_logic := '0';
begin

    for i in num'high downto num'low loop
        if(num(i) = '1') then
            ret := '1';
            exit;
        end if;
    end loop;
    return(ret);

end function;

function ZerosPresent(num: unsigned) return std_logic is
variable ret: std_logic := '0';
begin

    for i in num'high downto num'low loop
        if(num(i) = '0') then
            ret := '1';
            exit;
        end if;
    end loop;
    return(ret);

end function;


--  -----------------------
--  Declared constants
--  -----------------------
constant D_SIZ: integer := 1 + INT_SIZ + FRAC_SIZ; -- data word size
constant R_SIZ: integer := 1 + INT_SIZ + FRAC_SIZ; -- remainder size
constant Q_SIZ: integer := 1 + INT_SIZ + (2*FRAC_SIZ)+ 2; -- quotient size

constant D_MSB: integer := D_SIZ-1; -- data word msb bit position
constant R_MSB: integer := R_SIZ-1; -- remainder msb bit position

constant D_OVR: integer := D_SIZ; -- data word overflow bit position

constant Q_LSB: integer := 2; -- quotient lsb bit position
```

```vhdl
constant Q_MSB: integer := D_MSB+Q_LSB; -- quotient msb bit position


----------------------
--  Declared signals
----------------------
signal w1: signed(D_MSB downto 0) := (others=>'0');
signal w2: signed(D_MSB downto 0) := (others=>'0');

signal remainder: unsigned(R_MSB downto 0) := (others=>'0');
signal grs: unsigned(2 downto 0) := (others=>'0');

signal wq: signed(Q_SIZ-1 downto 0) := (others=>'0');
signal wq_prime: signed(Q_SIZ-1 downto 0) := (others=>'0');
signal q_prime:  signed(Q_SIZ downto 0) := (others=>'0');
signal q_prime_sel:  signed(Q_SIZ downto 0) := (others=>'0');
signal quo: signed(Q_SIZ downto 0) := (others=>'0');

signal a: unsigned(D_MSB downto 0) := (others=>'0');
signal b: unsigned(D_MSB downto 0) := (others=>'0');
signal sum: unsigned(D_MSB downto 0) := (others=>'0');

signal cnt: integer range 0 to Q_SIZ-1 := 0;
signal dividend_sign: std_logic := '0';
signal divisor_sign: std_logic := '0';
signal negate: std_logic := '0';

signal busy: std_logic := '0';
signal cor_cyc: std_logic := '0';
signal zero_remainder: std_logic := '0';
signal zero_det: std_logic := '0';
signal cor_needed: std_logic := '0';

signal presence_of_ones: std_logic := '0';
signal presence_of_zeros: std_logic := '0';
signal cor_presence_of_ones: std_logic := '0';
signal cor_presence_of_zeros: std_logic := '0';
signal sticky: std_logic := '0';

alias remainder_sign is remainder(remainder'high);
alias cor_remainder is sum;
```

```
-----------------------
-- Enumeration lists
-----------------------
type sm_def is
(
    RESET,
    START_DIV,
    DIV,
    DIV2,
    DIV3,
    ROUND,
    ROUND2
);
signal state: sm_def := RESET;

---------------------------------- module code ----------------------------
begin

--------------------------------------------
--
-- Signed Sequential Divider state machine
--
--------------------------------------------
process(rst,clk)
begin

    if(rst='1') then

            -- working registers
            w1  <= (others=>'0');
            w2  <= (others=>'0');
            remainder  <= (others=>'0');
            wq <= (others=>'0');
            quo <= (others=>'0');
            grs <= (others=>'0');

            -- local signals
            dividend_sign <= '0';
            divisor_sign <= '0';
            cnt <= 0;
            cor_cyc <= '0';
            zero_det <= '0';

            -- handshake signals
```

```vhdl
        busy <= '0';
        ovrflw <= '0';
        udrflw <= '0';

        -- states
        state <= RESET;

    elsif rising_edge(clk) then

        -- one clock signals
        cor_cyc <= '0';

        --
        --   state machine body
        --
        case state is
            -- reset state
            when RESET =>
                state <= START_DIV;
            --
            --   divider body
            --
            when START_DIV =>
                if(start = '1') then
                    w1 <= divisor;
                    divisor_sign <= divisor(divisor'high);
                    w2 <= dividend;
                    dividend_sign <= dividend(dividend'high);
                    -- prime remainder register with dividend sign
                    for i in remainder'high downto 0 loop
                        remainder(i) <= dividend(dividend'high);
                    end loop;
                    -- initialize variables
                    wq <= (others=>'0');
                    quo <= (others=>'0');
                    grs <= (others=>'0');
                    -- other
                    zero_det <= '0';
                    busy <= '1';
                    ovrflw <= '0';
                    udrflw <= '0';
                    cnt <= 0;
                    state <= DIV;
                end if;
```

```vhdl
-- division sequence
when DIV =>
    -- left-shift working dividend
    w2 <= w2(w2'high-1 downto 0)&'0';
    -- always update remainder with sum or difference
    remainder <= sum;
    -- set working quotient lsb bit while left-shifting all other bits
    wq <= wq(wq'high-1 downto wq'low)&negate;

    -- sequence counter
    if(cnt < Q_SIZ-1) then
        cnt <= cnt + 1;
        -- detect a remainder intermediate zero on all but last
        -- operational cycle
        zero_det <= zero_det or zero_remainder;
    else
        -- completed, start correction cycle
        cor_cyc <= '1';
        state <= DIV2;
    end if;
-- correction cycle
when DIV2 =>
    -- select between corrected and non-corrected working quotient
    if(cor_needed = '1') then
        quo <= q_prime_sel;
    else
        quo <= wq_prime&'1';
    end if;
    -- corrected sticky bit
    grs(0) <= sticky;
    state <= DIV3;
when DIV3 =>
    -- normalize quotient back into quotient register
    quo(D_OVR downto 0) <= quo(Q_MSB)&
                           quo(Q_MSB downto Q_LSB);
    state <= ROUND;
    -- set overflow flag based in sign
    for i in quo'high downto Q_MSB loop
        if(quo(i) /= quo(Q_MSB)) then
            ovrflw <= '1';
            busy <= '0';
            state <= START_DIV;
        end if;
    end loop;
```

```vhdl
                -- save guard and round bits from retired quotient bits
                grs(2 downto 1) <= unsigned(quo(1 downto 0));
            -- rounding for division
            when ROUND =>
                -- round result to nearest even number
                if (ROUNDING = '1') then
                    if(grs > 4 or (grs = 4 and quo(0) = '1')) then
                        -- round up
                        quo(D_OVR downto 0) <= quo(D_OVR downto 0) + 1;
                    else
                        null;-- truncate
                    end if;
                end if;
                state <= ROUND2;
            when ROUND2 =>
                -- check for overflow from rounding
                if(quo(D_OVR) /= quo(D_MSB)) then
                    ovrflw <= '1';
                -- data portion of operand equal to zero qualifies as an underflow
                elsif(quo(D_MSB downto 0) = 0) then
                    udrflw <= '1';
                end if;
                busy <= '0';
                state <= START_DIV;
            when others =>
                state <= RESET;
        end case;

    end if;

end process;


--------------------------
-- Primary adder circuit
--------------------------
-- add or subtract based on signs of remainder and divisor
negate <= (remainder_sign xnor divisor_sign);

-- remainder with next msb of dividend or remainder as is during correction cycle
a <= remainder(remainder'high-1 downto 0)&w2(w2'high) when cor_cyc = '0'
        else remainder;

-- subtract or add divisor
b <= unsigned((not(w1))+1) when negate = '1' else unsigned(w1);
```

```vhdl
-- sum and borrow outputs of adder
sum <= a + b;

-----------------------------
-- Quotient correction adder
-----------------------------
-- invert msb of working quotient
wq_prime <= (not wq(wq'high))&wq(wq'high-1 downto wq'low);

-- add 1
q_prime <= (wq_prime&'1') + 1;

-- select between +1 and -1 of working quotient
q_prime_sel <= q_prime when negate = '1' else wq_prime&'0';

-----------------------------
--   Correction cycle needed
-----------------------------
cor_needed <= '1' when (zero_det = '1' or (remainder_sign /= dividend_sign))
                  else '0';

-----------------------
-- Sticky decode logic
-----------------------
zero_remainder <= '1' when (remainder = 0 and w2 = 0) else '0';

-- check for ones and zeros without the sign bit
presence_of_ones <= OnesPresent(remainder(remainder'high-1 downto 0));
presence_of_zeros <= ZerosPresent(remainder(remainder'high-1 downto 0));
cor_presence_of_ones <= OnesPresent(cor_remainder(remainder'high-1 downto 0));
cor_presence_of_zeros <= ZerosPresent(cor_remainder(remainder'high-1 downto 0));

stky_gen:
process(dividend_sign,divisor_sign,cor_needed,cor_remainder,remainder,presence_of_ones,presence_of_zeros,cor_presence_of_ones,cor_presence_of_zeros)
begin

        --
        --   check remainder value
        --
        if(cor_needed = '1' and cor_remainder = 0) then
```

```vhdl
        sticky <= '0';
elsif(cor_needed = '0' and remainder = 0) then
        sticky <= '0';
else
    --
    -- (+) dividend
    --
    if(dividend_sign = '0') then
        -- (+) divisor
        if(divisor_sign = '0') then
            -- correction required
            if(cor_needed = '1') then
                -- corrected 1s present then sticky is 1 for positive quotient
                sticky <= cor_presence_of_ones;
            -- no correction required
            else
                -- 1s present then sticky is 1 for positive quotient
                sticky <= presence_of_ones;
            end if;
        -- (-) divisor
        else
            -- correction required
            if(cor_needed = '1') then
                -- corrected 1s present then sticky is 0 for negative quotient
                sticky <= not cor_presence_of_ones;
            -- no correction required
            else
                -- 1s present then sticky is 0 for negative quotient
                sticky <= not presence_of_ones;
            end if;
        end if;
    --
    -- (-) dividend
    --
    else
        -- (+) divisor
        if(divisor_sign = '0') then
            -- correction required
            if(cor_needed = '1') then
                -- corrected 0s present then sticky is 0 for negative quotient

                sticky <= not cor_presence_of_zeros;
            -- no correction required
            else
```

```vhdl
                    -- 0s present then sticky is 0 for negative quotient

                    sticky <= not presence_of_zeros;
                end if;
            -- (-) divisor
            else
                -- correction required
                if(cor_needed = '1') then
                    -- corrected 0s present then sticky is 1 for positive quotient

                    sticky <= cor_presence_of_zeros;
                -- no correction required
                else
                    -- 0s present then sticky is 1 for positive quotient
                    sticky <= presence_of_zeros;
                end if;
            end if;
        end if;
    end if;

end process;

-- output signals
cmplt <= (not start) and (not busy);
quotient <= quo(D_MSB downto 0);

end RTL;
```

7 SRT Division

SRT division is named after Sweeney, Robertson, and Tocher. There are numerous articles, papers, and books that offer very detailed explanations on the theory of operation and its limits. This chapter will not re-explain these items, but supply pertinent information for the reader to understand and design an SRT divider.

The intent of SRT is to achieve faster division. This is done by combining several techniques in order to limit, or eliminate where possible, carry propagation. First, instead of comparing the partial remainder to the divisor during each iterative cycle, it is compared to a constant, and the number of bits compared to that constant is reduced to just a few bits. Second, carry-save-adders are used instead of traditional adders. Not only that, but the partial remainder is maintained in carry-save form throughout the entire operation. Third, quotient bits are retired in the form of signed-digits (SD), which facilitates the use of on-the-fly or similar conversions from SD to two's complement.

Moreover, SRT can utilize higher radix dividers, which simply means more bits are retired during each iteration of the operation. Example, radix-2, radix-4, and radix-8 retire 1, 2, and 4 bits per clock cycle, respectively.

Limitations do exist. Namely that the SRT algorithm, in its purest form, operates on unsigned fractional operands. The good news is that the algorithm can be modified to support signed fixed-point operands while maintaining many of its performance benefits.

Section 7.1 will cover the basic concepts of SRT. Section 7.2 provides an SRT variant implementation for signed fixed-point in radix-2. Recommendations are provided for implementing a radix-4 design at the end.

7.1 SRT Basics

Basic SRT was designed to work with unsigned binary-fractional operands. The MSB being the sign bit with the next bit being the leading digit. This works well with floating-point because the leading 1 is implied and its position is always known: to the left of the implied binary point. Any repositioning of the leading digit can be done by shifting the implied 1 into the significand while adjusting the exponent accordingly. Resulting in a normalized operand for both the dividend and divisor.

$$S.1xxxxxxxxxxx$$

Using fixed-point is more difficult because the position of the leading digits in both operands is unknown.

The use of constants

The first concept in SRT is comparing the partial remainder to a constant instead of the divisor and only comparing a limited number of bits, in this case two. The constants are -1/2 and +1/2.

$$1.1 \quad \text{and} \quad 0.1$$

Initially, the divisor needs to be normalized so that its leading bit is in the 1/2 position (2^{-1}) to align itself with the bit weight of the constants. The dividend instead needs to be aligned to the 1/4 (2^{-2}) position, which becomes the starting value of the partial remainder.

$$S.1xxxxxxxxxxx \quad \text{Divisor}$$
$$S.S1xxxxxxxxxx \quad \text{Dividend}$$

During each cycle, the remainder is multiplied by 2, discarding the MSB and shifting in a zero into the LSB position, thus 2R. The upper twos bits of 2R are then compared against the -1/2 and +1/2 constants to determine the cycle's operation as well as the quotient bit's value.

Comparison Result	Quotient	Sub-operation
2R ≥ +1/2	plus1	New R = 2R - Divisor
2R < -1/2	minus 1	New R = 2R + Divisor
Neither	zero	New R = 2R

Note: *SRT signed division is slightly different and includes the signs of both the partial remainder and divisor.*

Each quotient bit is represent by a signed digit-set {-1,0,+1} . The advantage is that each bit contains carry content for easier conversion back to its required binary form.

Note: *During unsigned operations, if the final remainder is negative, a correction cycle is required on both the remainder and quotient. The divisor is added back to the remainder and 1 is subtracted from the quotient.*

Algorithm Flow

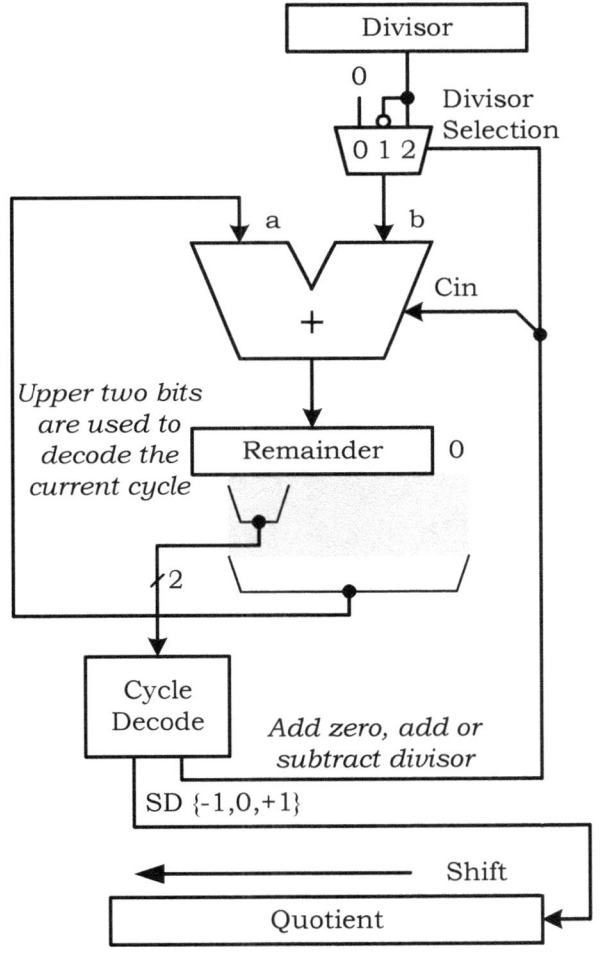

Figure 13

The use of On-The-Fly-Conversion

There are published on-the-fly-converters that convert SD {-1,0,+1} bits to two's complement as each quotient bit is retired, in particular the one by Ercegovac and Lang (1986). At the end of the division operation, there are two registers. One with the resulting quotient and the other with the resulting (quotient − 1). The selection between the two depends on the sign of the remainder, and whether correction is needed.

This only works with unsigned division, since signed division would require a (quotient + 1) alternative as well.

Limitations

- Unsigned division. Signed operations requires more logic.

- Conventional on-the-fly-conversion has only two alternatives instead of three.

7.2 SRT Radix-2 Division

As described in earlier chapters, SRT Radix-2 division only retires a single quotient bit per cycle, but it does so at a higher clock rate than other designs. The higher clock rate is achievable by reducing or eliminating the use of carry propagation adders in each stage and maintaining intermediate data in alternate forms. Most SRT algorithms support fractional or floating-point implementations only, but the algorithm in this section supports signed fixed-point with the integer and fraction portions configurable.

This last point greatly complicates the design's architecture because of the additional functions needed to normalize the operands to work within the SRT paradigm. It also makes the overall execution time data-dependent, meaning the number of clock cycles required depends on where the leading digits of both operands are.

Because of the sheer complexity of the design, a high-level overview of the architecture is provided first, followed by sections devoted to the details of sub-modules/functions.

While some designers may view the design as recondite, it does illustrate what is necessary to operate on fixed-point operands and reduce the effect of carry propagation throughout the design. Individual modules supporting particular aspects of the overall operation also serve as a venue to introduce the reader to intermediate number representations and their conversions. If anything, the reader should gain a new level of appreciation for high performance digital designers. The clock speeds provided at the end of the section completely justify the effort. Performance gains are substantial.

7.2.1 Architecture Overview

Figures 14 and 15 represent the SRT Radix-2 design. Both operands, the divisor and dividend, are input in their fixed-point two's complement format. The first step is to find the leading digit of each operand, whether the number is positive or negative, then reposition that leading bit into the 1/2 bit position, as required by the SRT algorithm. This is done by the Leading Digit Normalizer (LDN) module. The resulting direction shift and number bit positions shifted are saved for de-normalizing the quotient at the end of the division operation. This first step can also determine whether an overflow condition is imminent, and abort the operation immediately.

This design utilizes a standard carry-save-adder (CSA) instead of a traditional adder, maintaining the remainder in carry-save-form -- effectively eliminating carry propagation in this section. Using a CSA does, however, introduce side effects that must be compensated for. This issue is addressed in later sections.

The CSA supports the addition of three numbers: X, Y, and Z. Carry-in is injected into the unused bit position of the 'Y' input. This Allows the divisor to be added or subtracted (invert and add 1) in the standard two's complement fashion. The sum and carry outputs for each addition are fed-back, discarding the MSB, and updating the carry and sum registers, which combined represent the remainder in carry-save form.

Control of each cycle is provided by the *Cycle Decode ROM*, which is implemented as a synchronous block RAM and are available in most FPGA devices. On the rising edge of the clock, its inputs are latched into its internal address register, which immediately begin propagating the data value of that address to its outputs. Inputs include the same upper 4-bits that are also latched into both the carry and sum registers, the sign of the divisor, and cycle override control signal. Outputs include control of the divisor selection mux, carry into the CSA (of which there are two), and the value of the quotient digit for the current cycle.

The process begins by initializing the carry register to zero, the sum register with the normalized dividend, and the address register of the Cycle Decode ROM with the upper 4-bits of the normalized dividend.

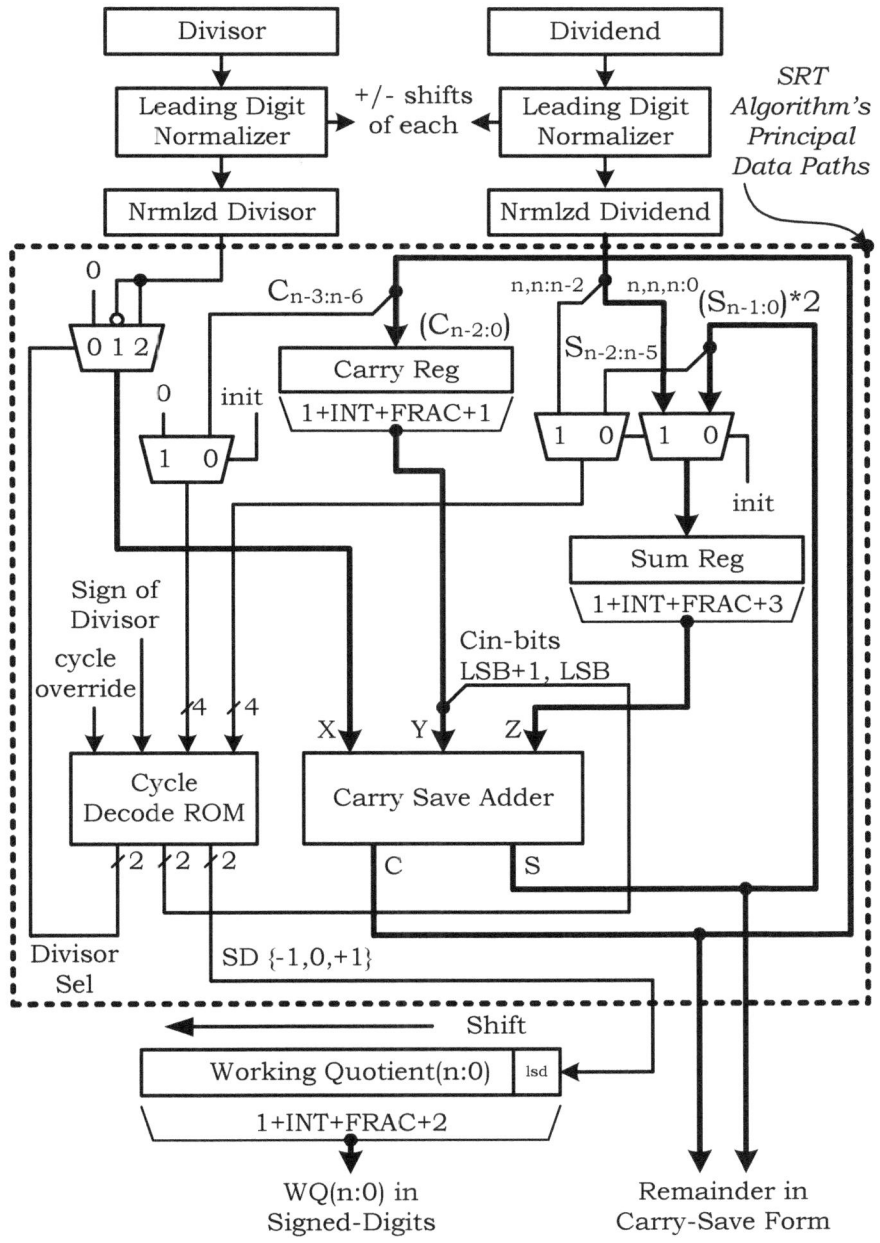

Figure 14

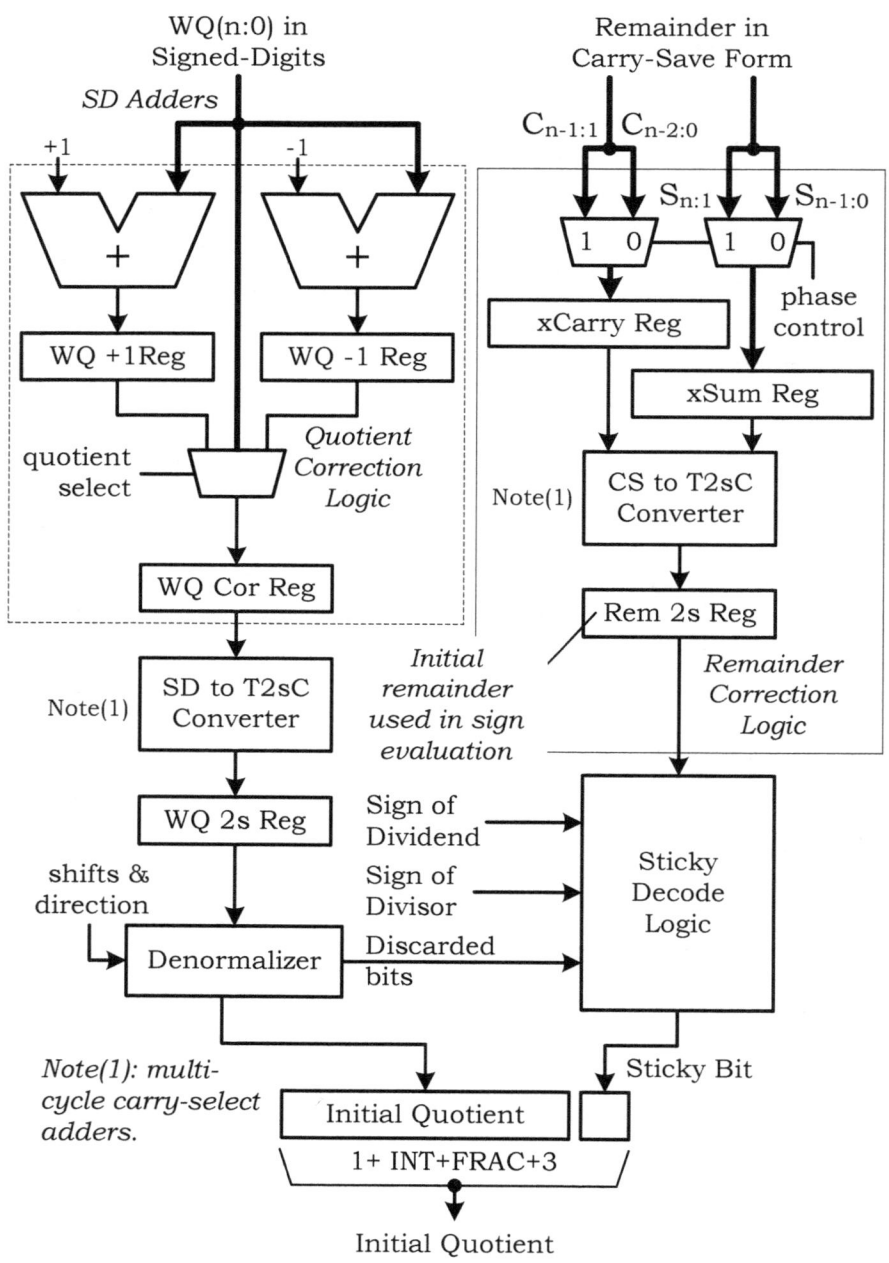

Figure 15

During each clock cycle the divisor (X input) is added to or subtracted from the values in the carry (Y input) and sum (Z input) registers. After the required number of iterations occur, which is the (operand size + 2 round bits), the working quotient register contains the resulting quotient, but in signed-digit (SD) format.

Similarly, the final remainder is in carry-save-form. Whether a correction is needed or not depends on if the signs of the dividend and remainder match or not. The remainder must first be converted from its carry-save-form into its two's complement form to make that determination.

While this conversion takes place, the (remainder + the divisor), and the (remainder − the divisor), are computed in the CSA loop and pipelined in right behind the initial remainder, so that any of the three may be used based on the results of the sign evaluation.

Concurrently, the working quotient has 1 added to it and 1 subtracted from it by two separate SD adders, making two corrected and one uncorrected working quotients available for selection depending on the sign evaluation of the remainder. SD adders are carry limited to adjacent bits only and therefore can complete in one clock cycle.

The selected remainder, corrected or not, along with the original operand signs are input to the sticky bit decode logic to determine the value of the sticky bit. The selected working quotient, corrected or not, must be converted from its SD form to its two's complement form, then denormalized based on the original normalization of the operands. Any discarded bits are also passed as an input to the sticky bit decode logic.

The initial quotient contains an extra high order bit in the event an overflow occurred during the correction or conversion stages. A high speed carry-select adder (not shown) is used in the final rounding operation.

Details are provided for each of these intermediate steps in subsequent sections.

7.2.2 Leading Digit Normalizer

The Leading Digit Normalizer (LDN) has two functions. First, it locates the leading digit in a fixed-point operand then shifts it so that its leading bit occupies the ½ weight position required by the SRT algorithm. The count and direction of the shift for both the dividend and divisor are retained to properly realign (denormalize) the quotient following the division operation. The number of shifts involved, relative to both operands, can also indicate whether the operation will overflow, thus allowing an early exit.

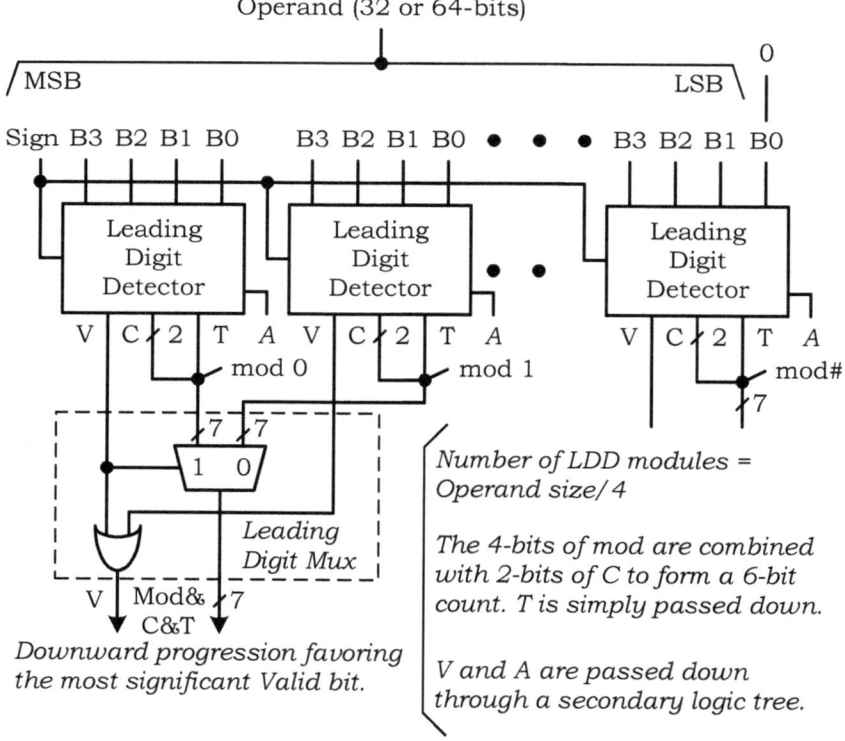

Figure 16

Locating the leading digit involves two separate logic trees with separate components. At the top of both trees are Leading Digit Detector (LDD) module, each evaluating four consecutive bits.

Collectively they span the operand from the MSB-1 down to its LSB.

The primary logic tree has a series of Leading Digit Mux (LDM) components that works very similar to leading zero detection circuits used throughout the industry. This logic gives precedence to the left-most mux input as the tree funnels down.

The LDD evaluates B3 down to B0 and outputs a 2-bit count 0 to 3, respectively. If the sign of the operand is positive the count (C) corresponds to the first '1' bit encounter, left to right. If operand is negative, the count (C) corresponds to the first '0', left to right. Likewise, the valid (V) bit is true if any bits are valid in the 4-bit group, '1' for positive and '0' for negative. 'T' indicates if there is a trailing-one following a zero within the 4-bits. The LDM gives precedence to the most significant LDD valid count, left to right, as V and C are funneled downward, as shown in Figures 16 and 17.

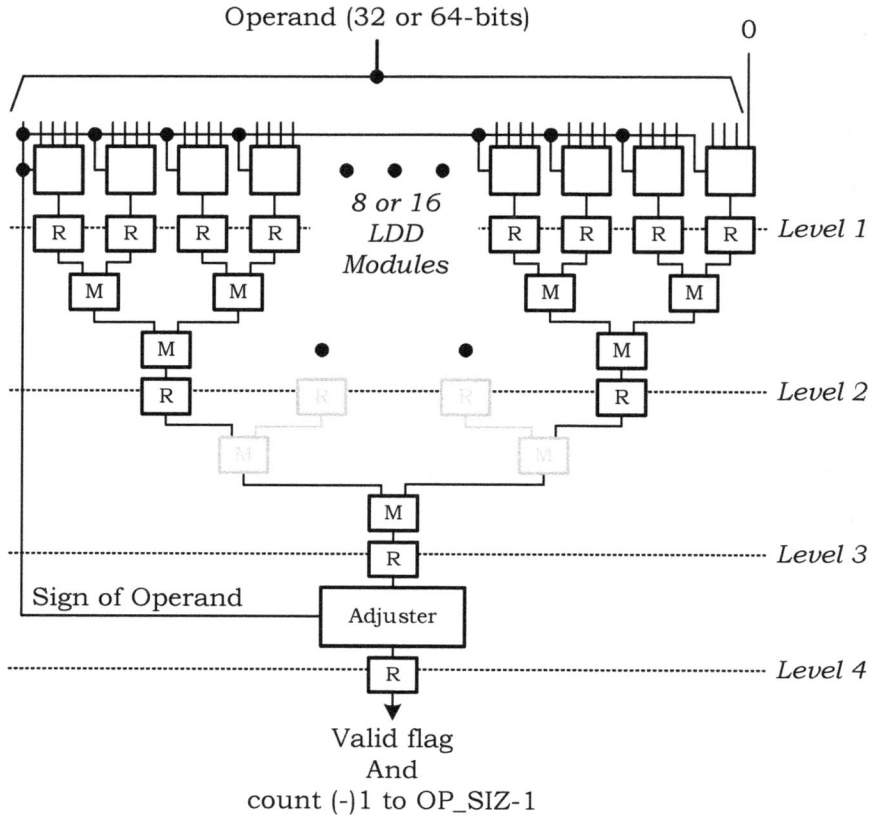

Figure 17

Note: *A '0' in the least significant bit in the last LDD allows proper detection of an operand with all ones.*

Figure 17 is heavily pipelined in order to minimize the levels of logic between stages. Registers are represented by 'R' and LDM muxes by 'M'. Between levels 3 and 4 is an adjuster circuit, which subtracts 1 from the resulting count under certain conditions. The rules are as follows:

- If the operand is negative, the leading digit is the first zero-bit encountered, left to right, from the sign bit.
- Unless, all lower bits are all zeros down to the LSB. In that case the leading digit is the last one-bit from the sign, which could include the sign bit itself

This last point necessitates the secondary logic tree, which properly funnels down the any-ones signal 'A' from each LDD Figure 16.

The second or background logic tree, uses both signals valid (V) and any-ones (A) from each LDD, as shown in Figure 18. LDD modules are grouped in fours, creating an OR of all V signals called *Next*. Signals *Next* and each 'A' signals are registered at level 1 just as all 'V' signals are.

Between levels 1 and 2, each 'A' signal is qualified by the 'V' signal of a higher order LDD, or *Next* from the previous group of four, *Previous Next*. This in essence qualifies the presence of ones of lower significances to the LDD that qualified the first zero in a negative number. Any trailing ones within the qualified LDD are qualified by its 'T' signal.

Figure 19 shows four grouping of four for a 64-bit operand, gating each next output to the previous inputs of lower significant stages.

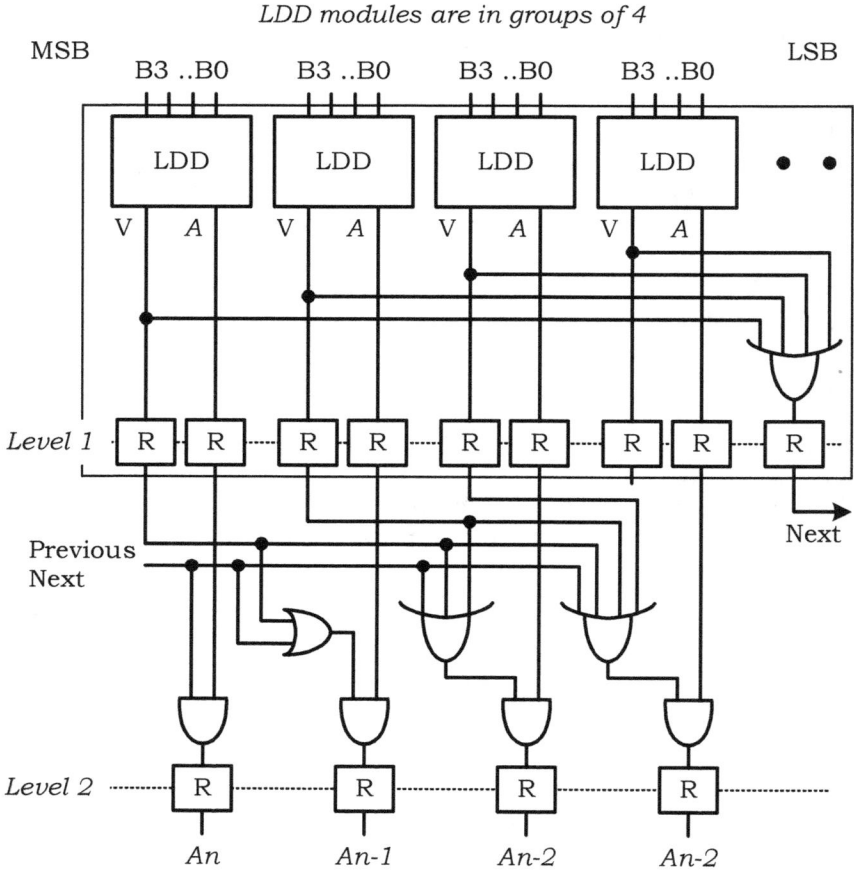

Figure 18

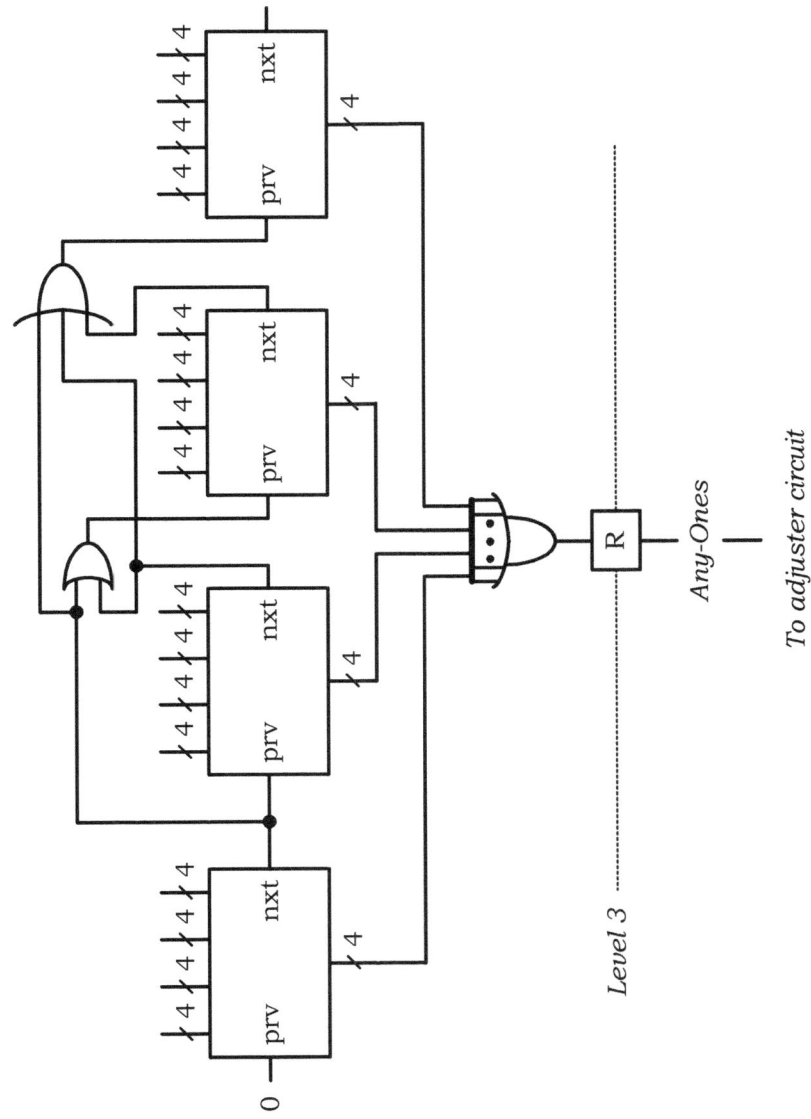

Figure 19

7.2.3 SRT Algorithm's Principal Data Paths

Accounted for here are those data paths within the dashed lines in Figure 14 called *SRT Algorithm's Principal Data Paths*.

Initialization

Shown below are all related registers and their corresponding values prior to the first cycle of the division algorithm. Some values are sign extended, which are explained in the notes that follow. The grayed areas are the bits that determine the first cycle's sub-operation.

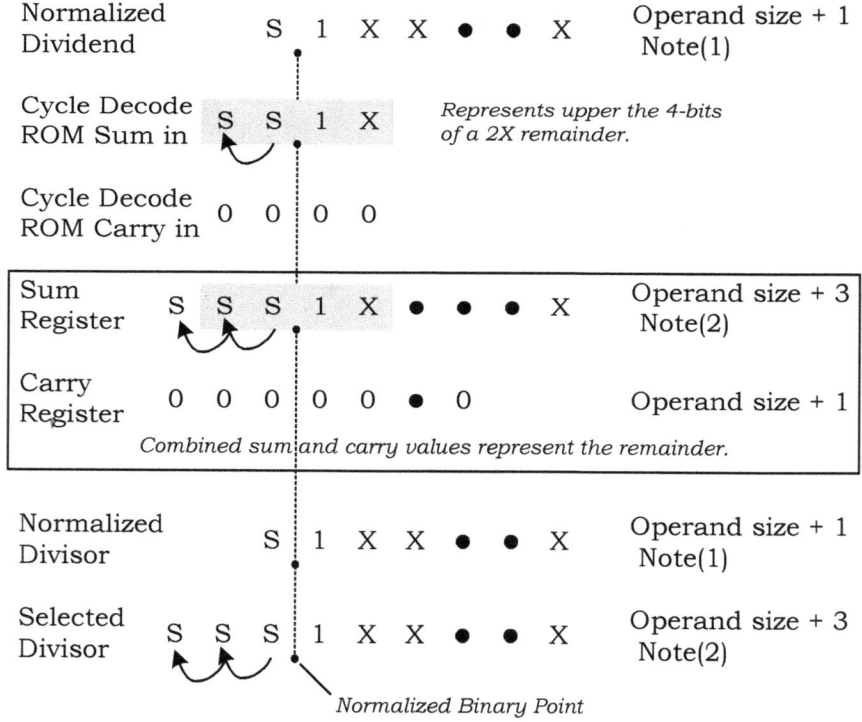

Note(1): *Both the normalized divisor and dividend are pre-extended one LSB bit in size. This is in the event that the operand was right shifted by one bit in the Leading Digit Normalizer module, in order to prevent the truncation of the LSB.*

Note(2): *The sum register is initialized to the value of the normalized dividend sign extended by two bits. This properly aligns the leading digit in the remainder for the first cycle and provides an extra MSB value of the remainder during phases 2 and 3 of the remainder correction cycles. Likewise, the selected dividend value is sign extended by two-bits to properly align with the dividend.*

Adder Circuits

The selected divisor together with the sum and carry registers, are input to a standard CSA's X, Y, and Z inputs, respectively. Since the carry of each bit propagates down instead of across and the sums of the combined carry and sum bits are not computed, they are maintained in carry-save form. The results of each cycle are simply placed back into the carry and sum registers, left shifted, and the MSB is discarded.

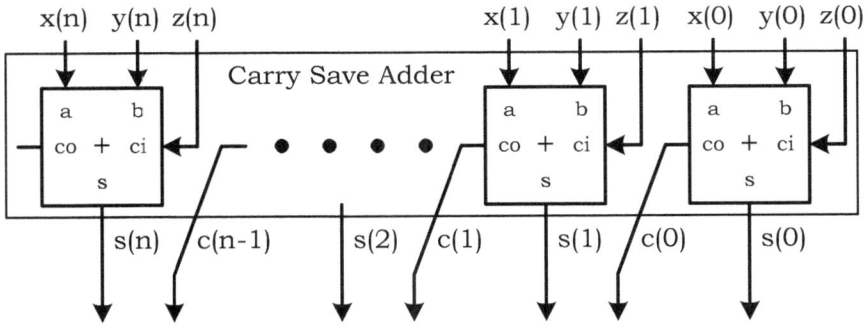

Figure 20

Figure 21 shows how each input is aligned relative to the other inputs and the location of the normalized binary point, shown by the dashed line. Each time sum and carry registers are updated, they are offset by zero, creating the 2X remainder, as required by the SRT algorithm. The lower two-bits of Y are not normally occupied. The lowest is used to insert a carry when the selected divisor is subtracted during normal cycles. The next higher bit position is used to insert carry during phases 2 and 3 of the remainder correction cycles.

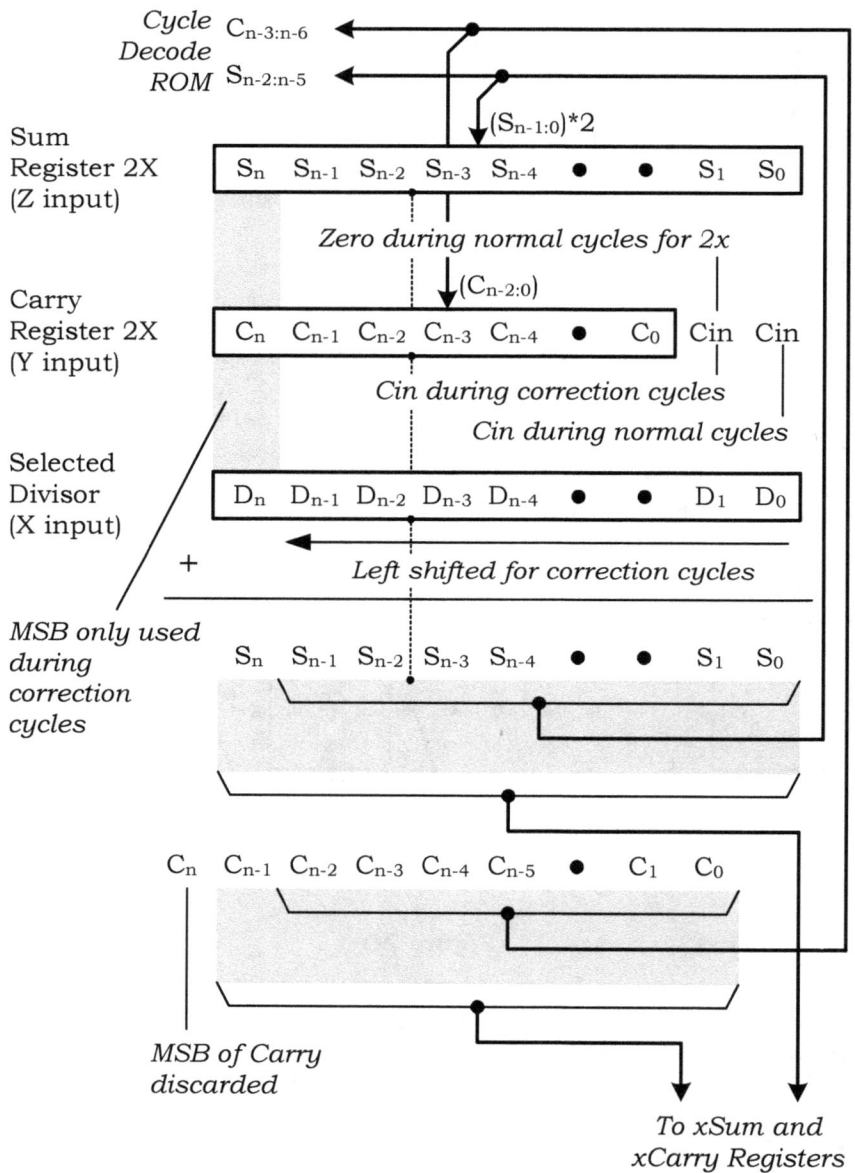

Figure 21

The new carry and sum resulting from each cycle are fed back into the carry and sum registers. The MSB of each is discarded and a zero is appended to the LSB, hence a left shift of the rest of the remainder. Four bits for the next cycle are also loaded into the address register of the *Cycle Decode ROM*. Notice that the extra MSB bits are discarded.

During the final cycle, the carry and sum values are also latched into the xcarry and xsum registers, minus their MSB bits. The next two cycles generate a remainder with the divisor added back and a remainder with the divisor subtracted from it. These are pipelined into the xcarry and xsum registers right behind the original remainder.

The original remainder is passed through as-is, but the MSB is discarded by the selection mux to the xCarry and xSum registers. The second step, referred to has phase 2 of the remainder correction cycle, adds the divisor back to the remainder. This is accomplished by left-shifting the divisor by one bit. This re-alignment nullifies the 2X effect of the carry and sum values. Phase 3 adds the inverted divisor while inserting a *Cin* of '1' into the LSB+1 position of the *Y* input. The LSB bits of the latter two phases are discarded in the selection mux to the xCarry and xSum registers.

During phases 2 and 3 the carry and sum registers are not updated again, because they retain the value of the original remainder.

Cycle Decode ROM

The Cycle Decode ROM (CDR) provides gating for the current cycle of the SRT algorithm and phases 2 and 3 of the remainder correction cycles. Shown in Figure 22 the CDR is implemented with an internal address register and a large lookup table. The inputs are latched in the address register, then data content at that address propagates out to control the next cycle.

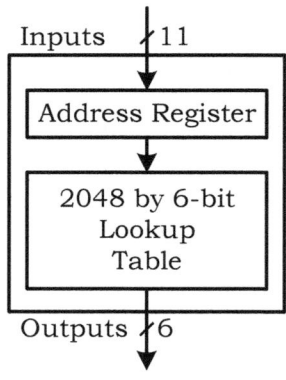

Figure 22

Note: *This component is easily inferred as a synchronous block RAM in FPGA devices with its lookup table content initialized during instantiation by a function.*

Inputs

- Carry and Sum (8-bits)

The new remainder value created from the CSA, in carry-save form, is used as an input, but only 4-bits each. Furthermore, they are evaluated in the lookup table as 2X, as required by the SRT algorithm.

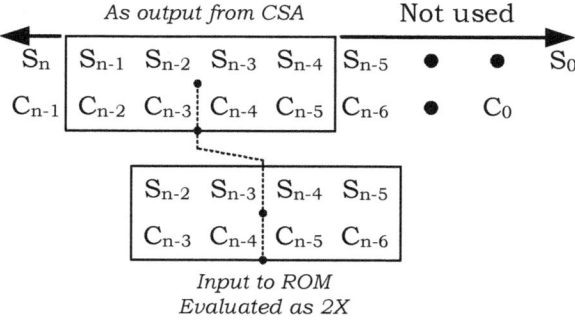

Figure 23

- Divisor Sign (1-bit)

The sign of the divisor changes how the carry and sum inputs are evaluated. The divisor sign is also compared against the remainder sign which is computed with the carry and sum inputs.

- Override (2-bits)

These signals override the effect of the carry and sum signals and force the CSA to either add or subtract the divisor to or from the remainder. This only occurs during phases 2 and 3 of the remainder correction cycles.

Outputs

- Divisor Selection (2-bits)

Selects either 0 or the divisor or an inverted divisor, supporting two's complement adding and subtraction of the divisor. When the inverted divisor is selected, the *Cin* associated with carry register, LSB position of the CSA *Y* input, is '1'.

- Quotient Digit (2-bits)

During each cycle, a new quotient digit, in SD form, is retired.

- Carry-in Bits (2-bits)

When the divisor is subtracted, *Cin* is inserted into the LSB of the Y input, and a zero into LSB+1. During the remainder corrections cycles, during a subtraction *Cin* is inserted into the LSB+1 position of the CSA *Y* input and the LSB a zero.

Lookup Table

As mentioned in an earlier section, because only the upper bits of the remainder in carry-save form are evaluated and the lower bits are ignored. There is inherent error because the entire remainder has not be converted to its final two's complement form. In order to minimize error the integer, ½ bit and ¼ bit positions are evaluated as well. Additional lower bits can be used if need be.

Override	Signs		Upper 4-bits	SD	Sub-operation
	Divisor	Remainder	Carry + Sum R(remainder)	Quotient	
00	+	X	≥ 0	{+1}	R = 2R - Divisor
00	+	X	≥ (-)1/2	{0}	R = 2R
00	+	X	< (-)1/2	{-1}	R = 2R + Divisor
00	-	X	≥ 0	{-1}	R = 2R + Divisor
00	-	X	≥ (-)1/2	{0}	R = 2R
00	-	X	< (-)1/2	{+1}	R = 2R - Divisor
10	X	X	X	X	R = R + 2(Divisor)
11	X	X	X	X	R = R - 2(Divisor)

Note: *R represents the remainder but in carry-save form.*

7.2.4 Carry-Save Form

The remainder value circulates within the SRT Algorithm in carry-save form. Its upper bits: the sign, integer bit, ½ bit and ¼ bit positions, are indirectly converted into a 4-bit two's complement number by the Cycle Decode ROM on every cycle.

The final remainder must be converted into a two's complement number in order to determine whether it needs to be corrected, along with the working quotient

Conversion is as simple as adding the carry bits (offset by 0), to the sum bits with a standard carry propagation adder. This, however, would nullify efforts to maintain a higher clock rate. Instead, a Carry-Select Adder is used. See section 7.2.6.

7.2.5 Signed Digit Representation

As each working quotient bit is retired, it is retired as a signed-digit. After the principal SRT algorithm has completed, a working quotient plus 1 and a working quotient minus 1 are computed, along with the value from the working quotient register.

Once the remainder has been converted into its two's complement form, one of the three working quotient values can be selected to be converted into its two's complement form.

Signed-Digits contain some carry information within each digit, allowing signed digits to be added to or subtracted from other signed-digit numbers extremely fast due to limiting carry propagation to adjacent bits only – the current bits and the previous bits.

There are twos steps. First, an intermediate sum (u) and carry (c) must be generated based on the rules shown in the table below. X and Y represent the two operands being added. Xi and Yi represent a bit from each at a specific index; whereas Xi-1 Yi-1 represent the lesser significant adjacent pair. Mirrored pairs {+1}{0} and {-1}{0} are duplicate, and {+1}{-1} along with its mirror, are invalid.

$x_i\ y_i$	{0}{0}	{0}{+1}	{0}{+1}	{0}{-1}	{0}{-1}	{+1}{+1}	{-1}{-1}
x_{i-1} y_{i-1}	n/a	neither is {-1}	Least one is {-1}	neither is {-1}	Least one is {-1}	n/a	n/a
u_i	{0}	{+1}	{0}	{0}	{-1}	{+1}	{-1}
c_i	{0}	{-1}	{+1}	{-1}	{+1}	{0}	{0}

The second step involves combining u and c to create a single signed-digit for each bit. The truth table is as follows:

c_i	u_i	sum
{0}	{0}	{0}
{0}	{-1}	{-1}
{-1}	{0}	{-1}
{0}	{+1}	{+1}
{+1}	{0}	{+1}
{+1}	{-1}	{0}
{-1}	{+1}	{0}
other	other	illegal

Converting a signed-digit number to its two's complement form requires a standard carry propagation adder. As with the remainder, a fast adder is used, but the conversion algorithm is as follows:

- First the signed-digit is split into two binary numbers, a high and low segment.
- Each bit in the high segment is set to 1 if the signed-digit is {-1}, otherwise 0. Each bit in the low segment is set to 1 if the signed-digit is {+1}, otherwise 0.
- Subtract the upper segment from the lower in two's complement fashion, invert and add 1.

7.2.6 Fast Carry-Select Adders

Conversion from both carry-save form of the remainder and the signed-digit representation of the working quotient into two's complement require a carry propagation adder. Using such adders on large operands would undo any performance gain in the design, so the use of a high performance fast adder is an imperative. Figure 24 represents such a design.

A series of smaller carry propagation adders are used instead, thus reducing the carry propagation time to the length of a small adder. Operand bits are evenly spread across these smaller adders. The tradeoff is that the conversion will take multiple clock cycles.

During the first clock cycle, the base segment adds the lower bits of the operands and registers the carry and sum. Concurrently, a series of small adders add the upper bits of the operands, generating a sum with the carry-in, one adder for a carry-in = 0 and one with carry-in = 1. During the second clock the registered carry from the first segment selects the results of the first upper segments by way of a data mux, both the sum and carry are registered. This continues through the chain of small adders.

> Note: *While the design provided evenly spreads the operand bits across the small adders, a design could have longer adder chains in the upper and final segments, thus reducing the number of clock cycles required.*

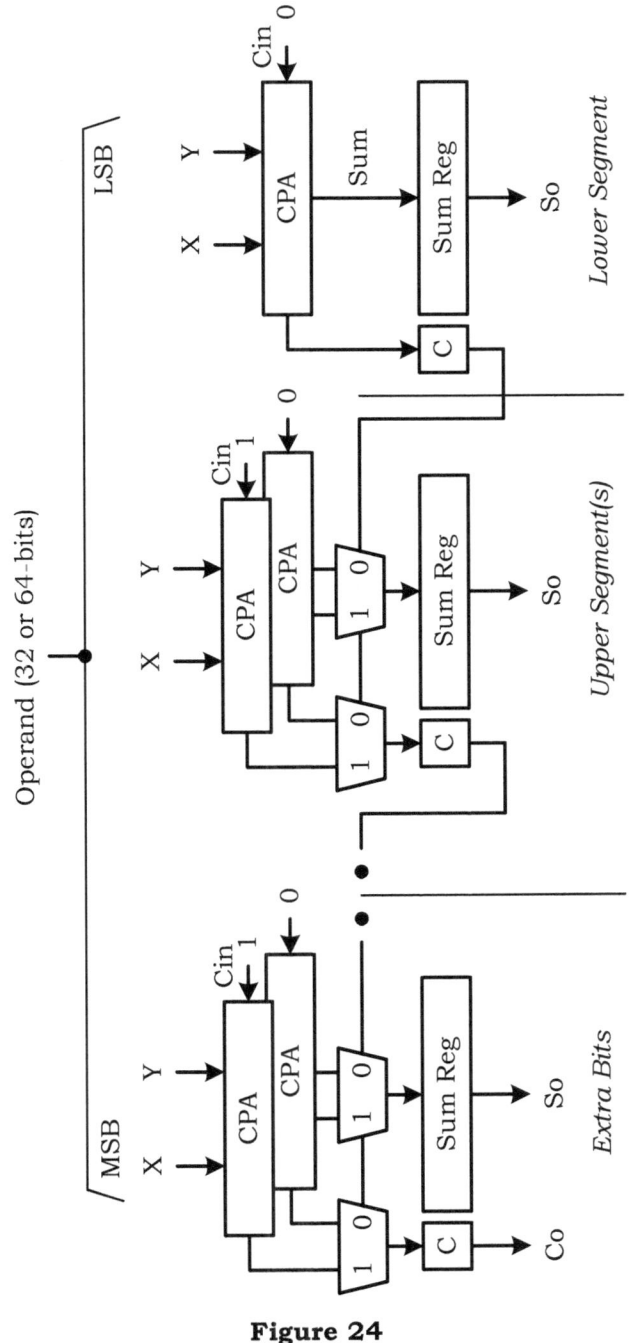

Figure 24

7.2.7 Denormalizer

The denormalizer is required because the design supports fixed-point operands. The resulting quotient is restored to the same fixed-point alignment as the dividend and divisor. The LDN modules, covered earlier, moves the leading-digit of both operands to the ½-bit position.

The range is from -1 to the (operand size − 1). A value -1 means the operand was shifted to the right by 1 bit, effectively assigning the leading bit with the sign-bit. The range of values from 0 to (operand size − 1) means the operand was shifted left by that many bit positions.

Immediately following the normalization of both operands, the number shifts required to reconstruction the final quotient is computed, as shown below. An early overflow occurs if the result is less than -1, and early underflow occurs if the result is greater than the operand size.

Operand size-2 + (dividend shifts − (number of fractional bits + number of rounding bits) - divisor shifts)

The (-2) accounts for the sign bit and ½-bit position, the number of fractional bits in the operands format (generic), and the number of round bits is 2 for guard and round (sticky is not included).

7.2.8 Correction, Rounding and Sticky Decode

The input to the denormalizer has already been corrected and so has its associated remainder in previous stages. The rules for correcting are the same as in previous chapters using signed division. They are:

- If the signs of the dividend and remainder do not match, both the quotient and remainder need correcting.
- If the signs of the remainder and divisor are equal
  Quotient = working quotient + 1
  Remainder = remainder − divisor
- If the signs of the remainder and divisor are not equal
  Quotient = working quotient − 1
  Remainder = remainder + divisor

The rules used in rounding the final quotient are also the same as in previous chapters. There are two additional bits in the quotient, guard and round, which are used in conjunction with sticky. The sticky bit, however, is slightly more complicated in this design.

In most cases, the denormalizer will right shift the working quotient which results in truncating those lower bits. Those bits are not discarded but are feed back into the sticky decode logic, providing the presence of additional ones or additional zeros.

Performance

The number of clock cycles required to complete the operation is data dependent. Normalizing both operands and denormalizing the quotient take additional clock cycles, thus a higher overhead burden.

In previous designs, the initial clocks cycles are used to detect an overflow, which is equal to the number of fractional bits in the fixed-point format. These cycles are not necessary in this design method because overflow can be detected during the normalization phase.

Optional Improvements

Significant time is spent normalizing the operands to the SRT requirement then returning the quotient to its original fixed-point format. Performance improvements in these circuits would have the greatest returns in reducing clock cycles.

Extending the design to support a higher radix, like radix-4, would reduce the number clock in the SRT algorithm by half. This would involve expanding the Cycle Decode ROM, using a CSA tree instead of a single CSA adder, retiring radix-4 signed digits {-3,-2,-1,0,+1,+2,+3} into the working quotient, employing radix-4 adders for working quotient correction, and then radix-4 to two's complement conversion. While this may seem overwhelming, it is not. Textbooks that cover radix-2 design also cover radix-4 design, information to expand the Cycle Decode ROM and CSA adder are present. Their operations are virtually identical, except that they use a higher radix signed digit. The concept of overlap is also possible, but requires a great deal of gate resources.

Example Design

The example state machine provided illustrates the *SRT2 Divider*.

- Both operands (divisor and dividend) and quotient are the same fixed-point format.
- Only two operand sizes are supported, 32 and 64-bit, but both have scalable Qm.n, integer and fractional bit lengths as generic parameters, defaulting to Q31.32, which is 64-bits.
- Signed two's complement numbers are supported for both the operands and quotient.
- Rounding can be enabled or disabled, but defaults to enabled; balanced round-to-nearest-even is used.
- There are additional sub modules.

As configured for Q31.32, test builds were run using Xilinx ISE. The following performance figures were reached with corresponding parts.

    Xilinx Spartan XC3s500e greater than 133MHz
    Xilinx Virtex XC5vlx30 greater than 304MHz

```vhdl
----------------------------------------------------
--
--    Srt2Divider.vhd
--
----------------------------------------------------
library IEEE;
use IEEE.std_logic_1164.all;
use IEEE.numeric_std.all;
-- user packages
use work.sd2_pkg.all;

use work.logic_pkg.all;

entity Srt2Divider is
--
--    Qmn fixed point format is used.
--
--              sign       binary point
--               |          |
--    format <s>(integer bits).(fractional bits)
--               _____/ _____/
--                  INT_SIZ      FRAC_SIZ
--
--    Note: only operands of 32 and 64-bit are supported!!!!
--    The combination of sign, integer, and fraction must equal
--    either 32 or 64 bits.
--
generic (INT_SIZ: integer range 0 to 63 := 31;
         FRAC_SIZ: integer range 0 to 63 := 32;
         ROUNDING: std_logic := '1');
port
(
    clk: in std_logic; -- system clock
    rst: in std_logic; -- system reset (must be synchronous)
    -- inputs
    start: in std_logic; -- start division
    divisor: in signed((1 + INT_SIZ + FRAC_SIZ)-1 downto 0);
    dividend: in signed((1 + INT_SIZ + FRAC_SIZ)-1 downto 0);
    -- ouputs
    cmplt: out std_logic; -- division complete
    ovrflw: out std_logic; -- overflow error
    udrflw: out std_logic; -- underflow error
    nrmerr: out std_logic; -- normalizing error
    quotient: out signed((1 + INT_SIZ + FRAC_SIZ)-1 downto 0)
```

);
end Srt2Divider;

architecture RTL of Srt2Divider is

-- Type and declared constants
----
constant OP_SIZ: integer := 1 + INT_SIZ + FRAC_SIZ; -- operand size
constant RD_SIZ: integer := 2; -- both guard and round bits included

-- COMPONENTS
component csla is
-- operand size must be greater or equal to partition size
generic (SIZ,PRTN_SIZ: integer);
port
(
   clk: in std_logic; -- system clock
   rst: in std_logic; -- system reset (must be synchronous)
   -- inputs
   wr: in std_logic;
   ci: in std_logic;
   x: in unsigned(SIZ-1 downto 0);
   y: in unsigned(SIZ-1 downto 0);
   -- outputs
   rdy: out std_logic;
   so: out unsigned(SIZ-1 downto 0);
   co: out std_logic
);
end component;

-- Declared signals

-- state machine signals
signal override: unsigned(1 downto 0) := (others=>'0');
signal init: std_logic := '0';

-- leading digit normalizer signals (operands extended 1 lsb bit)
signal dvsr_ldn_err: std_logic;
signal dvsr_ldn_rdy: std_logic;
signal dvsr_ldn_out: unsigned((OP_SIZ+1)-1 downto 0);
signal dvsr_ldn_shifts: integer range -1 to OP_SIZ-1;

```vhdl
signal dvnd_ldn_err: std_logic;
signal dvnd_ldn_rdy: std_logic;
signal dvnd_ldn_out: unsigned((OP_SIZ+1)-1 downto 0);
signal dvnd_ldn_shifts: integer range -1 to OP_SIZ-1;

alias dvsr_sign is dvsr_ldn_out(dvsr_ldn_out'high);
alias dvnd_sign is dvnd_ldn_out(dvnd_ldn_out'high);

-- sign extended operand sources
signal dvsr_src: unsigned((OP_SIZ+3)-1 downto 0) := (others=>'0');
signal dvnd_src: unsigned((OP_SIZ+3)-1 downto 0) := (others=>'0');

-- cycle decode rom signals
signal cdr_s: unsigned(3 downto 0) := (others=>'0');
signal cdr_c: unsigned(3 downto 0) := (others=>'0');
signal dvsr_sel: unsigned(1 downto 0) := (others=>'0');
signal cin: unsigned(1 downto 0) := (others=>'0');
signal sd: unsigned(1 downto 0) := (others=>'0');

-- sum and carry registers
signal sr: unsigned((OP_SIZ+3)-1 downto 0) := (others=>'0');
signal cr: unsigned((OP_SIZ+1)-1 downto 0) := (others=>'0');

-- carry save adder signals
signal x: unsigned((OP_SIZ+3)-1 downto 0) := (others=>'0');
signal y: unsigned((OP_SIZ+3)-1 downto 0) := (others=>'0');
signal z: unsigned((OP_SIZ+3)-1 downto 0) := (others=>'0');
signal s: unsigned((OP_SIZ+3)-1 downto 0) := (others=>'0');
signal c: unsigned((OP_SIZ+3)-1 downto 0) := (others=>'0');

-- remainder correction cycle registers and fast adder signals
signal xs: unsigned((OP_SIZ+2)-1 downto 0) := (others=>'0');
signal xc: unsigned((OP_SIZ+2)-1 downto 0) := (others=>'0');
signal r_2s: unsigned((OP_SIZ+2)-1 downto 0) := (others=>'0');
signal r_fa_start: std_logic := '0';
signal r_fa_rdy: std_logic := '0';

-- working quotient has an extra sign bit to detect early overflows, plus
-- 2 rounding bits, guard and round
signal wq: sd2_type((1+OP_SIZ+RD_SIZ)-1 downto 0) := (others=>"00");
signal cnt: natural range 0 to wq'high := 0;
signal wq_plus1: sd2_type(wq'high downto 0) := (others=>"00");
signal wq_minus1: sd2_type(wq'high downto 0) := (others=>"00");
signal wq_cor: sd2_type(wq'high downto 0) := (others=>"00");
```

```vhdl
signal wq_2s: signed(wq'high-1 downto 0) := (others=>'0');

-- quotient fast adder signals
signal q_fa_start: std_logic := '0';
signal q_fa_ci: std_logic := '0';
signal q_fa_x: unsigned(wq'high downto 0) := (others=>'0');
signal q_fa_y: unsigned(wq'high downto 0) := (others=>'0');
signal q_fa_rdy: std_logic := '0';
signal q_fa_so: unsigned(wq'high downto 0) := (others=>'0');

constant SD2_PLUS1:   sd2_type(wq'high downto 0) :=
T2sC_to_SD2(to_signed(1,wq'length));
constant SD2_MINUS1: sd2_type(wq'high downto 0) :=  T2sC_to_SD2(to_signed(-1,wq'length));
-- range of shift required for denormalizing
signal wq_shifts: integer range -OP_SIZ to OP_SIZ+2 := 0;

-- Denormalizer signals
signal start_dnm: std_logic;
signal dmn_rdy: std_logic := '0';
signal dmn_q: signed(wq_2s'high downto 0) := (others=>'0');
signal dis_zeros: std_logic := '0';
signal dis_ones: std_logic := '0';

-- final quotient and rounding signals
signal r_2s_ones: std_logic := '0';
signal r_2s_zeros: std_logic := '0';
signal sticky: std_logic := '0';
signal grs: unsigned(2 downto 0) := (others=>'0');
signal quo: unsigned(OP_SIZ-1 downto 0) := (others=>'0');

-- debug signals

-----------------------
--   Enumeration lists
-----------------------
type sm_def is
(
    RESET,
    START_DIV,
    NORMALIZE_OPS,
    CHK_FOR_OVRFLW,
    CHK_FOR_OVRFLW2,
    SRT_ALGO,
```

```vhdl
        REM_COR,
        REM_COR2,
        REM_COR3,
        FINAL_STEPS,
        FINAL_ROUND,
        FINAL_CHECK
);
signal state: sm_def := RESET;


---------------------------------- module code ----------------------------
begin


----------------------------------------
--
--    SRT Radix-2 Divider state machine
--
----------------------------------------
process(rst,clk)
begin

    if(rst='1') then

        -- working registers
        sr <= (others=>'0');
        cr <= (others=>'0');
        xs <= (others=>'0');
        xc <= (others=>'0');

        -- working quotient registers
        wq <= (others=>"00");
        wq_plus1 <= (others=>"00");
        wq_minus1 <= (others=>"00");
        wq_cor <= (others=>"00");
        wq_2s <= (others=>'0');

        -- misc
        dvsr_src <= (others=>'0');

        -- fast adder signals
        r_fa_start <= '0';
        q_fa_start <= '0';

        -- control
        override <= "00";
```

```vhdl
            cnt <= 0;
            init <= '0';

            -- denormalizing
            start_dnm <= '1';
            wq_shifts <= 0;

            -- quotient and rounding
            r_2s_ones <= '0';
            r_2s_zeros <= '0';
            grs <= (others=>'0');
            quo <= (others=>'0');

            -- handshake and error signals
            ovrflw <= '0';
            udrflw <= '0';
            nrmerr <= '0';

            -- states
            state <= RESET;

    elsif rising_edge(clk) then

            -- one clock signals
            r_fa_start <= '0';
            q_fa_start <= '0';
            start_dnm <= '0';

            -- continously pipelined signals
            wq_plus1 <= SD2_Adder(wq,SD2_PLUS1);
            wq_minus1 <= SD2_Adder(wq,SD2_MINUS1);
            r_2s_ones <= or_bits(r_2s);
            r_2s_zeros <= nand_bits(r_2s);


            --
            --  state machine body
            --
            case state is
                -- reset state
                when RESET =>
                    state <= START_DIV;

                --
                --  divider body
                --
```

```vhdl
when START_DIV =>
    -- wait for start command
    if(start = '1') then
        -- normalization has started with ldn modules
        ovrflw <= '0';
        udrflw <= '0';
        nrmerr <= '0';
        init <= '1';
        override <= "11";
        state <= NORMALIZE_OPS;
    end if;
when NORMALIZE_OPS =>
    -- wait for normalization to complete. normalized dividend and shift
    -- variables are accessed from the ldn ports, no need for
    -- extra copies. the divisor is maintained in state machine so that it
    -- can be left shifted during the remainder correction cycles.
    if(dvsr_ldn_rdy = '1' and dvnd_ldn_rdy = '1') then
        -- check for error
        if(dvsr_ldn_err = '1' or dvnd_ldn_err = '1') then
            nrmerr <= '1';
            state <= START_DIV;
        else
            override <= "00";
            state <= CHK_FOR_OVRFLW;
        end if;
    end if;
    -- normalized divisor sign extended 2-bits
    dvsr_src <= dvsr_sign&dvsr_sign&dvsr_ldn_out;

-- calculate wq denormalizing shifts, then check for imminent overflow
when CHK_FOR_OVRFLW =>
    -- the -2 accounts for the S.d bit positions
    wq_shifts <= (wq_2s'length-2) +
                 (dvnd_ldn_shifts - (FRAC_SIZ+RD_SIZ) –
                 dvsr_ldn_shifts);
    state <= CHK_FOR_OVRFLW2;
when CHK_FOR_OVRFLW2 =>
    -- beyond the sign position, however a -1 can be an overflow if
    -- all lower bits are not zero. this is determined prior to
    -- denormalizing.
    if(wq_shifts < -1) then
        ovrflw <= '1';
        state <= START_DIV;
    -- beyond the guard bit
```

```vhdl
        elsif(wq_shifts > OP_SIZ) then
            udrflw <= '1';
            state <= START_DIV;
        else
            state <= SRT_ALGO;
        end if;
        -- initialize registers for SRT algorithm
        init <= '0';
        cnt <= wq'high-2;
        sr <= dvnd_src;
        cr <= (others=>'0');
        wq <= (others=>"00");
--
--      SRT algorithm
--
    when SRT_ALGO =>
        -- 2x each input while discarding the msb of the last add
        sr <= s(s'high-1 downto 0)&'0';
        cr <= c(c'high-2 downto 0); -- lower bits set by ROM at Y inputs
        wq <= wq(wq'high-1 downto 0)&sd;
        -- select lower bits of remainder for correction pipeline
        xs <= s(s'high-1 downto 0);
        xc <= c(c'high-2 downto 0)&'0';

        -- loop here and let the cycle decode ROM do the rest
        if(cnt > 0) then
            cnt <= cnt - 1;
            -- start phase 2 of remainder correction
            if(cnt = 1) then
                override <= "01";    -- subtract divisor from remainder
            end if;
        -- execute remainder correction cycles
        else
            -- start phase 3 of remainder correction
            override <= "10";    -- add divsior back to remainder
            -- left shift divisor for phases 2 and 3 of remainder correction
            dvsr_src <= dvsr_src(dvsr_src'high-1 downto 0)&'0';
            -- start conversion of initial remainder in pipeline, which are
            -- the last inputs to xs and xc (above)
            r_fa_start <= '1';
            state <= REM_COR;
        end if;
```

```vhdl
--
-- Remainder and quotient correction/conversion
--
when REM_COR =>
    -- select upper bits of remainder (remainder - divisor), which
    -- is the second remainder in the pipeline.
    xs <= s(s'high downto 1);
    xc <= c(c'high-1 downto 1)&'0';
    state <= REM_COR2;
when REM_COR2 =>
    -- wait for initial remainder conversion from carry-save form
    -- to two's complement.
    if(r_fa_rdy = '1') then

        -- evaluate converted remainder and see if correction needed
        if(r_2s(r_2s'high) /= dvnd_sign) then
        --if(r_2s(r_2s'high) = '1') then
            -- correction required, WQ = WQ+1, R = R-Divisor
            if(r_2s(r_2s'high) = dvsr_sign) then
                wq_cor <= wq_plus1; -- select WQ+1
            -- correction required, WQ = WQ-1, R = R+Divisor
            else
                wq_cor <= wq_minus1; -- select WQ-1
                -- select upper bits of remainder (remainder + divisor),
                -- which is the third remainder in the pipeline. it should be
                -- lingering.
                xs <= s(x'high downto 1);
                xc <= c(x'high-1 downto 1)&'0';
            end if;
            -- start second remainder conversion with either
            -- (rem - divisor) or (rem + divisor). the using the latest xs and
            -- xc values.
            r_fa_start <= '1';
        -- no correction needed
        else
            wq_cor <= wq; -- select WQ as is, so is remainder
        end if;
        -- start conversion of working quotient from signed-digit to
        -- two's complement.
        q_fa_start <= '1';
        state <= REM_COR3;
    end if;

when REM_COR3 =>
```

```vhdl
        -- wait for final remainder and quotient conversions
        if(r_fa_rdy = '1' and q_fa_rdy = '1') then
            -- save converted value
            wq_2s <= signed(q_fa_so(q_fa_so'high-1 downto 0));
            -- check for overflow following quotient correction
            if(q_fa_so(q_fa_so'high) /= q_fa_so(q_fa_so'high-1)) then
                    ovrflw <= '1';
                    state <= START_DIV;
            else
                    -- check for overflow (unless maximum negative number)
                    if(wq_shifts = -1) then
                        -- postive and negative are handled differently
                        if(q_fa_so(q_fa_so'high-1) = '0') then
                            ovrflw <= '1';
                            state <= START_DIV;
                        -- exclude LSB
                        elsif(q_fa_so(q_fa_so'high-3 downto 1) /= 0) then
                            ovrflw <= '1';
                            state <= START_DIV;
                        else
                            start_dnm <= '1';
                            state <= FINAL_STEPS;
                        end if;
                    -- denormalize quotient
                    else
                        start_dnm <= '1';
                        state <= FINAL_STEPS;
                    end if;
            end if;
        end if;
--
--  Final steps
--
    when FINAL_STEPS =>
        -- wait for denormalizing of quotient
        if(dmn_rdy = '1') then
            state <= FINAL_ROUND;
        end if;
        -- capture quotient and roundind bits
        grs <= unsigned(dmn_q(1 downto 0))&sticky;
        quo <= unsigned(dmn_q(dmn_q'high downto 2));
    when FINAL_ROUND =>
        -- use fast adder for rounding
        if(ROUNDING = '1' and(grs > 4 or (grs = 4 and quo(0) = '1'))) then
```

```vhdl
                    -- use fast adder again    quo <= quo + 1;
                    q_fa_start <= '1';
                    state <= FINAL_CHECK;
                elsif(quo = 0) then
                    udrflw <= '1';
                    state <= START_DIV;
                else
                    state <= START_DIV;
                end if;
            when FINAL_CHECK =>
                -- wait for quotient rounding
                if(q_fa_rdy = '1') then
                    quo    <= q_fa_so(q_fa_so'high-3 downto 0);
                    -- final check for overflow or underflow
                    if(q_fa_so(q_fa_so'high downto q_fa_so'high-3) /= "000") and
                      (q_fa_so(q_fa_so'high downto q_fa_so'high-3) /= "111") then
                        ovrflw <= '1';
                    elsif(q_fa_so(q_fa_so'high-2 downto 0) = 0) then
                        udrflw <= '1';
                    end if;
                    state <= START_DIV;
                end if;

            when others =>
                state <= RESET;
        end case;

    end if;

end process;

-- output signals
cmplt <= '1' when (start = '0') and (state = START_DIV) else '0';
quotient <= signed(quo);


--------------------------------
--  Leading Digit Normalizers
--------------------------------
divisor_ldn_mod: entity work.ldn
generic map (OP_SIZ)
port map
(
    clk => clk,
    rst => rst,
```

```vhdl
    -- inputs
    op_wr => start,
    op_in => divisor,
    -- outputs
    op_err => dvsr_ldn_err,
    op_rdy => dvsr_ldn_rdy,
    op_out => dvsr_ldn_out,
    shifts => dvsr_ldn_shifts
);

dividend_ldn_mod: entity work. ldn
generic map (OP_SIZ)
port map
(
    clk => clk,
    rst => rst,
    -- inputs
    op_wr => start,
    op_in => dividend,
    -- outputs
    op_err => dvnd_ldn_err,
    op_rdy => dvnd_ldn_rdy,
    op_out => dvnd_ldn_out,
    shifts => dvnd_ldn_shifts
);
-- normalized dividend sign extended by 2-bits
dvnd_src <= dvnd_sign&dvnd_sign&dvnd_ldn_out;
----------------------
--   Cycle Decode ROM
----------------------
cdr_mod: entity work.cdr port map
(
    clk => clk,
    -- inputs
    s => cdr_s,
    c => cdr_c,
    divisor_sign => dvsr_src(dvsr_src'high),
    override => override,
    -- outputs
    divisor_sel => dvsr_sel,
    cin => cin,
    q => sd
);
-- cdr sum and carry muxes used for initial cycle
```

```vhdl
cdr_s <= dvnd_src(dvnd_src'high-1 downto dvnd_src'high-4) when init = '1'
    else s(s'high-2 downto s'high-5);
cdr_c <= (others=>'0') when init = '1' else c(c'high-3 downto c'high-6);


-------------------------------------
-- Carry Save Adder (SRT Algorithm)
-------------------------------------
-- divisor selection
x <= (not dvsr_src) when dvsr_sel = "01" else dvsr_src when dvsr_sel = "10"
    else (others=>'0');
-- sum and carry
y <= cr&cin; -- lower 2-bits are control by ROM
z <= sr;

csa_mod: entity work.csa generic map(OP_SIZ+3)
port map (x,y,z,s,c);


-------------------------------------
-- Remainder conversion fast adder
-------------------------------------
rem_csla_mod: entity work.csla
generic map (SIZ=>r_2s'length, PRTN_SIZ => 12)
port map
(
    clk => clk,
    rst => rst,
    -- inputs
    wr => r_fa_start,
    ci => '0',
    x => xs,
    y => xc,
    -- outputs
    rdy =>r_fa_rdy,
    so => r_2s,
    co => open
);


------------------------
-- Quotient fast adder
------------------------
quo_csla_mod: entity work.csla
generic map (SIZ=>wq_cor'length ,PRTN_SIZ => 12)
port map
(
```

```vhdl
        clk => clk,
        rst => rst,
        -- inputs
        wr => q_fa_start,
        ci => q_fa_ci,
        x => q_fa_x,
        y => q_fa_y,
        -- outputs
        rdy => q_fa_rdy,
        so => q_fa_so,
        co => open
);

-- input selection mux to fast adder
process(state,wq_cor,quo)
begin
    -- wait for fast adder to convert between SD and Twos
    -- complement,
    if(state = REM_COR3) then
            -- convert working quotient from SD to 2's complement.
            -- Algorithm: subtract negative SD component from positive SD,
            -- component regardless of encoding type.
        q_fa_ci <= '1';
        for i in wq_cor'range loop
            if(wq_cor(i) = SD_P) then
                q_fa_x(i) <= '1';
                q_fa_y(i) <= '1';
            elsif(wq_cor(i) = SD_N) then
                q_fa_x(i) <= '0';
                q_fa_y(i) <= '0';
            else
                q_fa_x(i) <= '0';
                q_fa_y(i) <= '1';
            end if;
        end loop;
    -- otherwise adder used for rounding quotient, must sign extend
    -- because adder is bigger than final quotient.
    else
        q_fa_ci <= '1';
        q_fa_x <=  quo(quo'high)&quo(quo'high)&quo(quo'high)&quo;
        q_fa_y <= (others=>'0');
    end if;

end process;
```

```vhdl
------------------
-- Denormalizer
------------------
dnm_mod: entity work.dnm
generic map (OP_SIZ,RD_SIZ)
port map
(
    clk => clk,
    rst => rst,
    -- inputs
    op_wr => start_dnm,
    op_in => wq_2s,
    op_shifts => wq_shifts,
    -- outputs
    op_rdy => dmn_rdy,
    op_out => dmn_q,
    -- discarded bits
    dis_zeros => dis_zeros,
    dis_ones => dis_ones
);


-------------------------
-- Sticky Decode Logic
-------------------------
process(wq_2s,r_2s,r_2s_ones,r_2s_zeros,dis_zeros,dis_ones)
begin

    -- if remainder is zero, sticky is zero
    if(r_2s_ones = '0') then
        sticky <= '0';
    -- if initial quotient is positive (sticky set to 1)
    elsif(wq_2s(wq_2s'high) = '0') then
        -- if the remainder is positive
        if(r_2s(r_2s'high) = '0') then
            -- ones in remainder or discarded ones
            if(r_2s_ones = '1' or dis_ones = '1') then
                sticky <= '1';
            else
                sticky <= '0';
            end if;
        -- if the remainder is negative
        else
            -- zeros in remainder or discarded zeros
```

```vhdl
                if(r_2s_zeros = '1' or dis_zeros = '1') then
                    sticky <= '1';
                else
                    sticky <= '0';
                end if;
            end if;
        -- if quotient is negative (sticky set to 0)
        else
            -- if the remainder is positive
            if(r_2s(r_2s'high) = '0') then
                -- ones in remainder or discarded ones
                if(r_2s_ones = '1' or dis_ones = '1') then
                    sticky <= '0';
                else
                    sticky <= '1';
                end if;
            -- if the remainder is negative
            else
                -- zeros in remainder or discarded zeros
                if(r_2s_zeros = '1' or dis_zeros = '1') then
                    sticky <= '0';
                else
                    sticky <= '1';
                end if;
            end if;
        end if;

end process;

end RTL;
```

--
-- sd_pkg.vhd (Signed-Digit Package, Radix-2)
--

```vhdl
library IEEE;
use IEEE.std_logic_1164.all;
use IEEE.numeric_std.all;
use STD.textio.all;
use IEEE.std_logic_textio.all;

-- package header
package sd2_pkg is
    -- types
    type sd2_type is array (natural range <>) of unsigned(1 downto 0);

    -- signed-digit constants
    constant SD_N: unsigned(1 downto 0) := "10"; -- {-1}
    constant SD_Z: unsigned(1 downto 0) := "00"; -- {0}
    constant SD_P: unsigned(1 downto 0) := "01"; -- {+1}
    -- combinations
    constant SD_Z_Z: unsigned(3 downto 0) := SD_Z&SD_Z;
    constant SD_Z_P: unsigned(3 downto 0) := SD_Z&SD_P;
    constant SD_Z_N: unsigned(3 downto 0) := SD_Z&SD_N;
    constant SD_P_Z: unsigned(3 downto 0) := SD_P&SD_Z;
    constant SD_P_P: unsigned(3 downto 0) := SD_P&SD_P;
    constant SD_P_N: unsigned(3 downto 0) := SD_P&SD_N;
    constant SD_N_Z: unsigned(3 downto 0) := SD_N&SD_Z;
    constant SD_N_P: unsigned(3 downto 0) := SD_N&SD_P;
    constant SD_N_N: unsigned(3 downto 0) := SD_N&SD_N;

    -- functions
    function SD2_to_T2sC(sd_num: sd2_type) return signed;
    function T2sC_to_SD2(num :signed) return sd2_type;
    function SD2_Adder(x,y: sd2_type) return sd2_type;

    -- procedures
    procedure report_sd2(sd_num: sd2_type);

end;
```

```vhdl
-- package body
package body sd2_pkg is
    -- converts signed-digit number to two's complement (simulation only)
    function SD2_to_T2sC(sd_num: sd2_type) return signed is
    variable l,h: unsigned(sd_num'high downto sd_num'low) := (others=>'0');
    variable tc: signed(sd_num'high downto sd_num'low) := (others=>'0');
    begin

        -- break apart sd into upper and lower halves (encoding independent)
        for i in sd_num'high downto sd_num'low loop
            case sd_num(i) is
                when SD_N =>
                    h(i) := '1';
                when SD_P =>
                    l(i) := '1';
                when others =>
                    null;
            end case;
        end loop;
        -- subtract high from low in two's complement fashion
        tc := signed(l + ((not h)+1));
        return(tc);
    end function;

    -- converts two's complement number to signed-digit   (simulation only)
    function T2sC_to_SD2(num :signed) return sd2_type is
    variable ci,co: std_logic := '0';
    variable sd: sd2_type(num'length-1 downto 0) := (others=>SD_Z);
    variable eval: unsigned(2 downto 0) := (others=>'0');
    begin

        for i in num'low to num'high loop

            if(i < num'high) then
                eval := num(i+1)&num(i)&ci;
                case eval is
                    when "000" => sd(i) := SD_Z; co := '0';
                    when "001" => sd(i) := SD_P; co := '0';
                    when "010" => sd(i) := SD_P; co := '0';
                    when "011" => sd(i) := SD_Z; co := '1';
                    when "100" => sd(i) := SD_Z; co := '0';
                    when "101" => sd(i) := SD_N; co := '1';
                    when "110" => sd(i) := SD_N; co := '1';
                    when "111" => sd(i) := SD_Z; co := '1';
```

```vhdl
                    when others => null;
                end case;
            else
                eval := '0'&num(i)&ci;
                case eval is
                    when "000" => sd(i) := SD_Z; co := '0';
                    when "001" => sd(i) := SD_P; co := '0';
                    when "010" => sd(i) := SD_P; co := '0';
                    when "011" => sd(i) := SD_Z; co := '1';
                    when others => null;
                end case;
            end if;
            -- forward to next digit
            ci := co;

    end loop;

    return(sd);
end function;

-- adds two signed-digit numbers (synthesizable)
function SD2_Adder(x,y: sd2_type) return sd2_type is
variable u,c,sum: sd2_type(x'high downto x'low) := (others=>SD_Z);
variable eval: unsigned(3 downto 0) := (others=>'0');
begin

    -- generate intermediate sums (u) and carries (c)
    for i in x'high downto x'low loop
        eval := x(i)&y(i);
        case eval is
            --when SD_Z_Z =>
            when "0000" =>
                c(i) := SD_Z;
                u(i) := SD_Z;
            --when SD_Z_P | SD_P_Z =>
            when "0001" | "0100" =>
                if(i = x'low) then
                    c(i) := SD_P;
                    u(i) := SD_N;
                elsif(x(i-1) /= SD_N and y(i-1) /= SD_N) then
                    c(i) := SD_P;
                    u(i) := SD_N;
                else
                    c(i) :=  SD_Z;
```

```vhdl
            u(i) := SD_P;
        end if;
    --when SD_Z_N | SD_N_Z =>
    when "0010" | "1000" =>
        if(i = x'low) then
            c(i) := SD_Z;
            u(i) := SD_N;
        elsif(x(i-1) /= SD_N and y(i-1) /= SD_N) then
            c(i) := SD_Z;
            u(i) := SD_N;
        else
            c(i) :=  SD_N;
            u(i) := SD_P;
        end if;
    --when SD_P_P =>
    when "0101" =>
        c(i) := SD_P;
        u(i) := SD_Z;
    --when SD_N_N =>
    when "1010" =>
        c(i) := SD_N;
        u(i) := SD_Z;
    when others => null;
    end case;

end loop;
-- combine both u and c  for final sum
for i in sum'high downto sum'low loop

    if(i > sum'low) then
        eval := c(i-1)&u(i);
        case eval is
            --when SD_Z_Z => sum(i) := SD_Z;
            when "0000" => sum(i) := SD_Z;
            --when SD_Z_N | SD_N_Z => sum(i) := SD_N;
            when "0010" | "1000" => sum(i) := SD_N;
            --when SD_Z_P | SD_P_Z => sum(i) := SD_P;
            when "0001" | "0100" => sum(i) := SD_P;
            --when SD_P_N | SD_N_P => sum(i) := SD_Z;
            when "0110" | "1001" => sum(i) := SD_Z;
            when others =>  sum(i) := "XX"; -- illegal
        end case;
    else
        sum(i) := u(i); -- no carry in with lsb
```

```
        end if;

    end loop;

    return(sum);

end function;

-- print signed-digit number (simulation only)
procedure report_sd2(sd_num: sd2_type) is
variable L: line;
begin
    for i in sd_num'high downto sd_num'low loop
        case sd_num(i) is
            when SD_Z =>    write(L,string'("0"));
            when SD_P =>    write(L,string'("1"));
            when SD_N =>    write(L,string'("N"));
            when others =>  write(L,string'("X"));
        end case;
    end loop;
    writeline(output,L);
end procedure;

end;
```

```vhdl
----------------------------------------------------------
--
--    logic_pkg.vhd
--
----------------------------------------------------------
library IEEE;
use IEEE.std_logic_1164.all;
use IEEE.numeric_std.all;

--  package header
package logic_pkg is

    function or_bits(a: unsigned) return std_logic;
    function nand_bits(a: unsigned) return std_logic;
end;

--  package body
package body logic_pkg is

    -- ORing all inputs. Any input high produces a high output
    function or_bits(a: unsigned) return std_logic is
    variable r: std_logic := '0';
    begin
        for i in a'low to a'high loop
            if(a(i) = '1') then
                r := '1';
            end if;
        end loop;
        return(r);
    end function;

    -- NANDing all inputs. Any input low produces a high output
    function nand_bits(a: unsigned) return std_logic is
    variable r: std_logic := '0';
    begin
        for i in a'low to a'high loop
            if(a(i) = '0') then
                r := '1';
            end if;
        end loop;
        return(r);
    end function;

end;
```

```vhdl
---------------------------------------------------------
--
--   Ldn.vhd (Leading Digit Normalizer)
--
--   The input operand is normalized to the SRT requirement
--   that the leading digit be positioned to 1/2 bit position
--   with its sign extended to the integer position.
--
--   Operand     S.xxx1xxxxx
--               S.1xxxxxxxx
--
--   The count returned indicates the number of shifts required
--   to reposition the leading digit, (-)1 to 63. (-)1 is a right
--   shift, 1 to 63 is a left shift.
--
--   The leading digit for positive operands is the first 1
--   encountered. The first zero encountered for negative
--   numbers, provided that there are trailing ones. If not,
--   the last one before the zero is the leading digit.
--
--   The returned, or normalized operand is 1-bit larger. This
--   is in the event a right-shift was required, which would
--   result in truncating the LSB.
---------------------------------------------------------
library IEEE;
use IEEE.std_logic_1164.all;
use IEEE.numeric_std.all;
use work.logic_pkg.all;

entity Ldn is
-- Only operands of 32 and 64-bits are supported
generic (OP_SIZ: integer := 64);
port
(
    clk: in std_logic; -- system clock
    rst: in std_logic;  -- system reset (must be synchronous)
    -- inputs
    op_wr: in std_logic;
    op_in: in signed(OP_SIZ-1 downto 0);
    -- outputs
    op_err: out std_logic;
    op_rdy: out std_logic;
    op_out: out unsigned(OP_SIZ downto 0); -- 1 bit larger (right shift)
    shifts: out integer range -1 to OP_SIZ-1 -- (-)1 to (+)31 or 63
```

```vhdl
);
end Ldn;

architecture RTL of Ldn is

-- leading digit detector signals and constants
type cnt_array is array (natural range <>) of unsigned(5 downto 0);
constant START_BIT: natural := OP_SIZ-2; -- sign bit excluded
constant NUM_LDD: natural := OP_SIZ/4; -- number of ldd modules
constant NUM_MUXES_2A: natural := NUM_LDD/2; -- number of muxes for 2A
-- number of muxes 2B
constant NUM_MUXES_2B: natural := NUM_MUXES_2A/2;
-- number of muxes for 3A
constant NUM_MUXES_3A: natural := NUM_MUXES_2B/2;

signal v: unsigned(0 to NUM_LDD-1) := (others=>'0');
signal rv: unsigned(0 to NUM_LDD-1) := (others=>'0');
signal c: cnt_array(0 to NUM_LDD-1) := (others=>(others=>'0'));
signal rc: cnt_array(0 to NUM_LDD-1) := (others=>(others=>'0'));
signal t: unsigned(0 to NUM_LDD-1) := (others=>'0');
signal rt: unsigned(0 to NUM_LDD-1) := (others=>'0');
signal a: unsigned(0 to NUM_LDD-1) := (others=>'0');
signal ra: unsigned(0 to NUM_LDD-1) := (others=>'0');
signal nxt: unsigned(0 to (NUM_LDD/4)-1) := (others=>'0');
signal rnxt: unsigned(0 to (NUM_LDD/4)-1) := (others=>'0');

signal mv_2a: unsigned(0 to NUM_MUXES_2A-1) := (others=>'0');
signal mc_2a: cnt_array(0 to NUM_MUXES_2A-1) := (others=>(others=>'0'));
signal mt_2a: unsigned(0 to NUM_MUXES_2A-1) := (others=>'0');
signal mv_2b: unsigned(0 to NUM_MUXES_2B-1) := (others=>'0');
signal mc_2b: cnt_array(0 to NUM_MUXES_2B-1) := (others=>(others=>'0'));
signal mt_2b: unsigned(0 to NUM_MUXES_2B-1) := (others=>'0');
signal rv_2: unsigned(0 to NUM_MUXES_2B-1) := (others=>'0');
signal rc_2: cnt_array(0 to NUM_MUXES_2B-1) := (others=>(others=>'0'));
signal rt_2: unsigned(0 to NUM_MUXES_2B-1) := (others=>'0');
signal da_2: unsigned(0 to NUM_LDD-1) := (others=>'0');
signal rda_2: unsigned(0 to NUM_LDD-1) := (others=>'0');

signal mv_3a: unsigned(0 to NUM_MUXES_3A-1) := (others=>'0');
signal mc_3a: cnt_array(0 to NUM_MUXES_3A-1) := (others=>(others=>'0'));
signal mt_3a: unsigned(0 to NUM_MUXES_3A-1) := (others=>'0');
signal rv_3: std_logic := '0';
signal rc_3: unsigned(5 downto 0) := (others=>'0');
signal rt_3: std_logic := '0';
```

```vhdl
signal ra_3: std_logic := '0';
signal nc: signed(6 downto 0) := (others=>'0');
signal nv: std_logic := '0';

-- normalizer signals
signal op: unsigned(OP_SIZ-1 downto 0)  := (others=>'0');
signal wop: unsigned(OP_SIZ downto 0)  := (others=>'0'); -- working operand 1 bit larger
signal cnt: natural range 0 to 4 := 0;
signal err: std_logic := '0';
signal rdy: std_logic := '0';

type sm_def is
(
    RESET,
    INPUT_OP,
    WAIT_FOR_C,
    RT_SHFT_1,
    LF_SHFT_32,
    LF_SHFT_32_2,
    LF_SHFT_16,
    LF_SHFT_8,
    LF_SHFT_4,
    LF_SHFT_2,
    LF_SHFT_1,
    FINISHED
);
signal state: sm_def := RESET;

---------------------------------- module code ----------------------------
begin


---------------------------------------------
--
--   Leading Digit Normailizer state machine
--
---------------------------------------------
process(rst,clk)
begin

    if(rst='1') then

        op <= (others=>'0');
```

```vhdl
            wop     <= (others=>'0');
            cnt <= 0;

            err <= '0';
            rdy <= '0';
            op_out <= (others=>'0');
            shifts <= 0;

            state <= RESET;

    elsif rising_edge(clk) then

        case state is
            -- reset state
            when RESET =>
                state <= INPUT_OP;
            --
            -- state machine body
            --
            when INPUT_OP =>
                -- accept operand, locator logic runs in parallel
                if(op_wr = '1') then
                    op <= unsigned(op_in);
                    cnt <= 4; -- time required to locate leading digit
                    -- clear flags
                    err <= '0';
                    rdy <= '0';
                    state <= WAIT_FOR_C;
                end if;
            when WAIT_FOR_C =>
                -- load working operand   (lsb set to zero)
                wop <= op&'0';
                -- wait for leading digit count to complete
                if(cnt > 0) then
                    cnt <= cnt - 1;
                elsif(nv = '1') then
                    -- shift
                    if(nc(6) = '1') then
                        state <= RT_SHFT_1;
                    elsif(nc(5) = '1') then -- start with left shift 32-bits
                        state <= LF_SHFT_32;
                    elsif(nc(4) = '1') then -- start with left shift 16-bits
                        state <= LF_SHFT_16;
                    elsif(nc(3) = '1') then -- start with left shift 8-bits
```

```vhdl
            state <= LF_SHFT_8;
        elsif(nc(2) = '1') then -- start with left shift 4-bits
            state <= LF_SHFT_4;
        elsif(nc(1) = '1') then -- start with left shift 2-bits

            state <= LF_SHFT_2;
        elsif(nc(0) = '1') then  -- start with left shift 1-bit
            state <= LF_SHFT_1;
        else
            state <= FINISHED; -- no shifts needed
        end if;
    else
        err <= '1';
        state <= FINISHED;
    end if;
--
-- barrel shifting, 32,16,8,4,2,1
--
when RT_SHFT_1 => -- right shift 1-bit
    wop <= wop(wop'high)&wop(wop'high downto 1);
    rdy <= '1';
    state <= FINISHED;
when LF_SHFT_32 => -- left shift 16-bits two times
    wop(wop'high downto 1) <= wop(wop'high downto 1) sll 16;
    state <= LF_SHFT_32_2;
when LF_SHFT_32_2 =>
    wop(wop'high downto 1) <= wop(wop'high downto 1) sll 16;
    -- decide on next step
    if(nc(4) = '1') then
        state <= LF_SHFT_16;
    elsif(nc(3) = '1') then
        state <= LF_SHFT_8;
    elsif(nc(2) = '1') then
        state <= LF_SHFT_4;
    elsif(nc(1) = '1') then
        state <= LF_SHFT_2;
    elsif(nc(0) = '1') then
        state <= LF_SHFT_1;
    else
        state <= FINISHED;
    end if;
when LF_SHFT_16 => -- left shift 16-bits
    wop(wop'high downto 1) <= wop(wop'high downto 1) sll 16;
```

```vhdl
    -- decide on next step
    if(nc(3) = '1') then
        state <= LF_SHFT_8;
    elsif(nc(2) = '1') then
        state <= LF_SHFT_4;
    elsif(nc(1) = '1') then
        state <= LF_SHFT_2;
    elsif(nc(0) = '1') then
        state <= LF_SHFT_1;
    else
        state <= FINISHED;
    end if;
when LF_SHFT_8 => -- left shift 8-bits
    wop(wop'high downto 1) <= wop(wop'high downto 1) sll 8;
    -- decide on next step
    if(nc(2) = '1') then
        state <= LF_SHFT_4;
    elsif(nc(1) = '1') then
        state <= LF_SHFT_2;
    elsif(nc(0) = '1') then
        state <= LF_SHFT_1;
    else
        state <= FINISHED;
    end if;
when LF_SHFT_4 => -- left shift 4-bits
    wop(wop'high downto 1) <= wop(wop'high downto 1) sll 4;
    -- decide on next step
    if(nc(1) = '1') then
        state <= LF_SHFT_2;
    elsif(nc(0) = '1') then
        state <= LF_SHFT_1;
    else
        state <= FINISHED;
    end if;
when LF_SHFT_2 => -- left shift 2-bits
    wop(wop'high downto 1) <= wop(wop'high downto 1) sll 2;
    if(nc(0) = '1') then
        state <= LF_SHFT_1;
    else
        state <= FINISHED;
    end if;
when LF_SHFT_1 => -- left shift 1-bit
    wop(wop'high downto 1) <= wop(wop'high downto 1) sll 1;
    state <= FINISHED;
```

```vhdl
                when FINISHED =>
                    rdy <= '1';
                    shifts <= TO_INTEGER(nc);
                    op_out <= wop;
                    state <= INPUT_OP;

                when others =>
                    state <= RESET;
            end case;

        end if;

    end process;

    -- gated flags here
    op_err <= err when op_wr = '0' else '0';
    op_rdy <= rdy when op_wr = '0' else '0';

    -------------------------------------------------------
    --
    --  Locate leading digit through progressive mux tree,
    --  precedence to the left most bit. A secondary path
    --  is used to retain the presence of trailing ones.
    --
    -------------------------------------------------------
    gen_4bit_mods: for x in 0 to 0 generate
    begin
        --
        -- level one. valid, count, trailing ones, and any zeros are
        -- registered from ldd modules. the 'next' signal is an OR
        -- of 4 consecutive valid signals.
        --
        gen_ldd_mods: for i in 0 to (NUM_LDD)-1 generate
        begin
            -- assign upper bits of count based on module number
            c(i)(5 downto 2) <= TO_UNSIGNED(i,4);
            -- create the ldd modules required for the operand size
            gen_ldd_u: if i < (NUM_LDD)-1 generate
                ldd_upr: entity work.ldd port map
                (
                    polarity => op(op'high),
                    b => op((START_BIT - (i*4)) downto (START_BIT - (i*4))-3),
                    v => v(i),
                    c => c(i)(1 downto 0),
```

```vhdl
            t => t(i),
            a => a(i)
        );
    end generate;
    gen_ldd_l: if i = (NUM_LDD)-1 generate
        ldd_lwr: entity work.ldd port map
        (
            polarity => op(op'high),
            b(3 downto 1) => op(2 downto 0),
            b(0) => '0', -- additional zero is for negative numbers
            v => v(i),
            c => c(i)(1 downto 0),
            t => t(i),
            a => a(i)
        );
    end generate;
    -- OR four ldd module's V ouputs to create a next signal (i=15 not used)
    gen_nxt0: if i = 3 generate
        nxt(0) <= v(0) or v(1) or v(2) or v(3);
    end generate;
    gen_nxt1: if i = 7 generate
        nxt(1) <= v(4) or v(5) or v(6) or v(7);
    end generate;
    gen_nxt2: if i = 11 generate
        nxt(2) <= v(8) or v(9) or v(10) or v(11);
    end generate;

end generate;
-- register all signals representing level 1
process(rst,clk)
begin

    if(rst='1') then
        rv <= (others=>'0');
        rc <= (others=>(others=>'0'));
        rt <= (others=>'0');
        ra <= (others=>'0');
        rnxt <= (others=>'0');
    elsif rising_edge(clk) then
        rv <= v;
        rc <= c;
        rt <= t;
        ra <= a;
        rnxt <= nxt;
```

```
        end if;

end process;
--
-- Level two. diverges into two paths. first, two stages of muxed
-- counts and trailing ones are selected by the left most V signal
-- and are then registered. the second path enables any-ones
-- from lower significant ldd modules. these constitute trailing
-- ones beyond the qualifying ldd.
--

-- muxed counts and ldd trailing ones stage A
gen_2a_ldm_mux: for i in 0 to NUM_MUXES_2A-1 generate
begin
        gen_ldm: entity work.ldm port map
        (
            vl => rv(i*2),
            cl => rc(i*2),
            tl => rt(i*2),
            vr => rv((i*2)+1),
            cr => rc((i*2)+1),
            tr => rt((i*2)+1),
            v => mv_2a(i),
            c => mc_2a(i),
            t => mt_2a(i)
        );

end generate;
-- muxed counts and ldd trailing ones of stage B
gen_2b_ldm_mux: for i in 0 to NUM_MUXES_2B-1 generate
begin
        gen_ldm: entity work.ldm port map
        (
            vl => mv_2a(i*2),
            cl => mc_2a(i*2),
            tl => mt_2a(i*2),
            vr => mv_2a((i*2)+1),
            cr => mc_2a((i*2)+1),
            tr => mt_2a((i*2)+1),
            v => mv_2b(i),
            c => mc_2b(i),
            t => mt_2b(i)
        );
```

```
end generate;

-- enable any-ones external to qualifying ldd
gen_ax: for i in 0 to (NUM_LDD)-1 generate
    -- any-ones decode
    gen_ax3: if(i = 3) generate
        da_2(0) <= '0'; -- can never quailfy
        da_2(1) <= ra(1) and rv(0);
        da_2(2) <= ra(2) and (rv(0) or rv(1));
        da_2(3) <= ra(3) and (rv(0) or rv(1) or rv(2));
    end generate;

    gen_ax7: if(i = 7) generate
        da_2(4) <= ra(4) and rnxt(0); -- previous
        da_2(5) <= ra(5) and (rv(4) or rnxt(0));
        da_2(6) <= ra(6) and (rv(4) or rv(5) or rnxt(0));
        da_2(7) <= ra(7) and (rv(4) or rv(5) or rv(6) or rnxt(0));
    end generate;

    gen_ax11: if(i = 11) generate
    signal prvs: std_logic := '0';
    begin
        prvs <= rnxt(0) or rnxt(1); -- previous
        da_2(8) <= ra(8) and prvs;
        da_2(9) <= ra(9) and (rv(8) or prvs) ;
        da_2(10) <= ra(10) and (rv(8) or rv(9) or prvs);
        da_2(11) <= ra(11) and (rv(8) or rv(9) or rv(10) or prvs);
    end generate;

    gen_ax15: if(i = 15) generate
    signal prvs: std_logic := '0';
    begin
        prvs <= rnxt(0) or rnxt(1) or rnxt(2); -- previous
        da_2(12) <= ra(12) and prvs;
        da_2(13) <= ra(13) and (rv(12) or prvs);
        da_2(14) <= ra(14) and (rv(12) or rv(13) or prvs);
        da_2(15) <= ra(15) and (rv(12) or rv(13) or rv(14) or prvs);

    end generate;

end generate;

-- register muxed counts and trailing zeros, as well as decoded
-- any-ones
```

```
process(rst,clk)
begin

    if(rst='1') then
        rv_2 <= (others=>'0');
        rc_2 <= (others=>(others=>'0'));
        rt_2 <= (others=>'0');
        rda_2 <= (others=>'0');
    elsif rising_edge(clk) then
        rv_2 <= mv_2b;
        rc_2 <= mc_2b;
        rt_2 <= mt_2b;
        rda_2 <= da_2;
    end if;

end process;


--
--  Level three. two levels, a and b, of muxed counts and trailing ones
--  are registered. separately, the "any-ones" bits are all ORed together.
--
gen_3a_ldm_mux: for i in 0 to NUM_MUXES_3A-1 generate
begin
        gen_ldm: entity work.ldm port map
        (
            vl => rv_2(i*2),
            cl => rc_2(i*2),
            tl => rt_2(i*2),
            vr => rv_2((i*2)+1),
            cr => rc_2((i*2)+1),
            tr => rt_2((i*2)+1),
            v => mv_3a(i),
            c => mc_3a(i),
            t => mt_3a(i)
        );

end generate;
-- 32-bit requires no further muxing, whereas 64 requires two
gen_3b_ldm32_mux: if OP_SIZ = 32 generate

    process(rst,clk)
    begin

        if(rst='1') then
```

```vhdl
                rv_3 <= '0';
                rc_3 <= (others=>'0');
                rt_3 <= '0';
            elsif rising_edge(clk) then
                rv_3 <= mv_3a(0);
                rc_3 <= mc_3a(0);
                rt_3 <= mt_3a(0);
            end if;

    end process;

end generate;

gen_3b_ldm64_mux: if OP_SIZ = 64 generate
signal mv_3b: std_logic := '0';
signal mc_3b: unsigned(5 downto 0) := (others=>'0');
signal mt_3b: std_logic := '0';
begin

        gen_ldm: entity work.ldm port map
        (
            vl => mv_3a(0),
            cl => mc_3a(0),
            tl => mt_3a(0),
            vr => mv_3a(1),
            cr => mc_3a(1),
            tr => mt_3a(1),
            v => mv_3b,
            c => mc_3b,
            t => mt_3b
        );

        process(rst,clk)
        begin

            if(rst='1') then
                rv_3 <= '0';
                rc_3 <= (others=>'0');
                rt_3 <= '0';
            elsif rising_edge(clk) then
                rv_3 <= mv_3b;
                rc_3 <= mc_3b;
                rt_3 <= mt_3b;
            end if;
```

```vhdl
        end process;

    end generate;
    -- OR and register all the "any-ones" bits
    process(rst,clk)
    begin

        if(rst='1') then
            ra_3 <= '0';
        elsif rising_edge(clk) then
            ra_3 <= or_bits(rda_2);
        end if;

    end process;

end generate;

--
--   level four. final decode to account for exceptions
--
process(rst,clk)
begin

    if(rst='1') then
        nv <= '0';
        nc <= (others=>'0');
    elsif rising_edge(clk) then
        -- pass valid signal down
        nv <= rv_3;
        -- positive operand
        if(op(op'high) = '0') then
            nc <= signed('0'& rc_3);
        -- negative operand
        else
            -- if no trailing ones, last 1 before first zero is leading-bit
            if((rc_3 = OP_SIZ-1) or (rt_3 = '0' and ra_3 = '0')) then
                -- 7-bit adder
                nc <= signed('0'&rc_3 - 1);
            -- if trailing ones, first zero is leading-bit
            else
                nc <= signed('0'& rc_3);
            end if;
        end if;
```

    end if;

end process;

end RTL;

```vhdl
---------------------------------------------------------
--
--  Ldm.vhd (6-bit Leading Digit Mux)
--
--  Works in conjunction with the Ldd circuit, selecting
--  between two ldd values based on the valid bits. The
--  left most valid has priority.
--
--  Trailing ones associated with the selected count are
--  also included in the mux.
---------------------------------------------------------
library IEEE;
use IEEE.std_logic_1164.all;
use IEEE.numeric_std.all;

entity Ldm is
port
(
    -- inputs
    vl: in std_logic; -- left count valid
    cl: in unsigned(5 downto 0); -- left count
    tl: in std_logic; -- left trailing ones
    vr: in std_logic; -- right count valid
    cr: in unsigned(5 downto 0); -- right count
    tr: in std_logic; -- right trailing ones
    -- outputs
    v: out std_logic; -- either count valid
    c: out unsigned(5 downto 0); -- selected count
    t: out std_logic -- selected trailing ones
);
end Ldm;

architecture RTL of Ldm is

----------------------------------- module code ----------------------------
begin

v <= vl or vr;
c <= cl when vl = '1' else cr;
t <= tl when vl = '1' else tr;

end RTL;
```

```vhdl
--------------------------------------------------------
--
--    Ldd.vhd (4-bit Leading Digit Detector)
--
--    Locates the leading digit, 1 or 0, within a 4-bit field
--    from left to right and assigns a count of 0-3, with an
--    extra bit indicating whether the count is valid.
--
--------------------------------------------------------
library IEEE;
use IEEE.std_logic_1164.all;
use IEEE.numeric_std.all;

entity Ldd is
port
(
    -- inputs
    polarity: in std_logic; -- 0 = detects leading one, 1 = detects leading zero
    b: in unsigned(3 downto 0); -- bits
    -- outputs
    v: out std_logic := '0'; -- valid count, leading bit present
    c: out unsigned(1 downto 0); -- leading digit count
    t: out std_logic; -- trailing ones
    a: out std_logic -- any ones
);
end Ldd;

architecture RTL of Ldd is

---------------------------------- module code ----------------------------

begin

-- valid
v <= '1' when ((polarity = '0' and b /= 0) or (polarity = '1' and (not b) /= 0)) else '0';

process(polarity,b)
begin
        -- positive operands
        if(polarity = '0') then
            if(b(3) = '1') then
                c <= "00";
            elsif(b(2) = '1') then
                c <= "01";
```

```vhdl
            elsif(b(1) = '1') then
                c <= "10";
            elsif(b(0) = '1') then
                c <= "11";
            else
                c <= "00";
            end if;
        -- negative operands
        else
            if(b(3) = '0') then
                c <= "00";
            elsif(b(2) = '0') then
                c <= "01";
            elsif(b(1) = '0') then
                c <= "10";
            elsif(b(0) = '0') then
                c <= "11";
            else
                c <= "00";
            end if;
        end if;
end process;
-- a one trailing a zero, left to right
process(b)
begin
    case b is
        when "0000" => t <= '0';
        when "0001" => t <= '1';
        when "0010" => t <= '1';
        when "0011" => t <= '1';
        when "0100" => t <= '1';
        when "0101" => t <= '1';
        when "0110" => t <= '1';
        when "0111" => t <= '1';
        when "1000" => t <= '0';
        when "1001" => t <= '1';
        when "1010" => t <= '1';
        when "1011" => t <= '1';
        when "1100" => t <= '0';
        when "1101" => t <= '1';
        when "1110" => t <= '0';
        when others => t <= '0';
    end case;
```

end process;
-- any ones present
a <= '1' when b /= 0 else '0';

end RTL;

--
-- dnm.vhd Denormalizer
--
-- Returns number from normalized state to fixed-point,
-- as well flags indicating if any zeros or ones were
-- truncated by right shifting.
----
library IEEE;
use IEEE.std_logic_1164.all;
use IEEE.numeric_std.all;
use work.logic_pkg.all;

entity dnm is
generic(OP_SIZ,RD_SIZ: integer);
port
(
    clk: in std_logic; -- system clock
    rst: in std_logic; -- system reset (must be synchronous)
    -- inputs
    op_wr: in std_logic;
    op_in: in signed((OP_SIZ+RD_SIZ)-1 downto 0);
    op_shifts: in integer range -OP_SIZ to OP_SIZ;
    -- outputs
    op_rdy: out std_logic;
    op_out: out signed((OP_SIZ+RD_SIZ)-1 downto 0);
    -- discarded bits (used as input to sticky generation logic)
    dis_zeros: out std_logic;
    dis_ones: out std_logic
);
end dnm;

architecture RTL of dnm is

-- denormalizer signals
signal op: signed(op_in'high downto 0) := (others=>'0');
signal wop: signed(op_in'high downto 0) := (others=>'0');
signal rdy: std_logic := '0';
signal shifts: signed(7 downto 0) := (others=>'0');

signal zeros: unsigned(8 downto 0) := (others=>'0');
signal ones: unsigned(8 downto 0) := (others=>'0');

```vhdl
type sm_def is
(
    RESET,
    INPUT_OP,
    EVAL,
    LF_SHFT_1,
    RT_SHFT_64,
    RT_SHFT_64_2,
    RT_SHFT_32,
    RT_SHFT_32_2,
    RT_SHFT_16,
    RT_SHFT_8,
    RT_SHFT_4,
    RT_SHFT_2,
    RT_SHFT_1,
    FINISHED
);
signal state: sm_def := RESET;
---------------------------------- module code ----------------------------

begin

------------------------------
--
--    Denomalizer state machine
--
------------------------------
process(rst,clk)
begin

    if(rst='1') then

        op <= (others=>'0');
        shifts <= (others=>'0');
        rdy <= '0';
        op_out <= (others=>'0');

        zeros <= (others=>'0');
        ones <= (others=>'0');

        dis_zeros <= '0';
        dis_ones <= '0';

        state <= RESET;
```

```vhdl
elsif rising_edge(clk) then

    case state is
        -- reset state
        when RESET =>
            state <= INPUT_OP;
        --
        --   state machine body
        --
        when INPUT_OP =>
        -- accept operand
            if(op_wr = '1') then
                op <= op_in;
                shifts <= to_signed(op_shifts,shifts'length);
                rdy <= '0';
                zeros <= (others=>'0');
                ones <= (others=>'0');
                dis_zeros <= '0';
                dis_ones <= '0';
                state <= EVAL;
            end if;
        when EVAL =>
            wop <= op;
            if(shifts(7) = '1') then
                state <= LF_SHFT_1;
            elsif(shifts(6) = '1') then -- start with right shift 64-bits
                state <= RT_SHFT_64;
            elsif(shifts(5) = '1') then -- start with right shift 32-bits
                state <= RT_SHFT_32;
            elsif(shifts(4) = '1') then -- start with right shift 16-bits
                state <= RT_SHFT_16;
            elsif(shifts(3) = '1') then -- start with right shift 8-bits
                state <= RT_SHFT_8;
            elsif(shifts(2) = '1') then -- start with right shift 4-bits
                state <= RT_SHFT_4;
            elsif(shifts(1) = '1') then -- start with right shift 2-bits
                state <= RT_SHFT_2;
            elsif(shifts(0) = '1') then   -- start with right shift 1-bit
                state <= RT_SHFT_1;
            else
                state <= FINISHED; -- no shifts needed
            end if;
```

```vhdl
--
--   barrel shifting: 64,32,16,8,4,2,1
--
when LF_SHFT_1 => -- left shift 1-bit
    wop <= wop sll 1;
    state <= FINISHED;
when RT_SHFT_64 => -- right shift 16-bits four times
    wop <= shift_right(wop,16);
    zeros(0) <= nand_bits(unsigned(wop(15 downto 0)));
    ones(0) <= or_bits(unsigned(wop(15 downto 0)));
    state <= RT_SHFT_64_2;
when RT_SHFT_64_2 => -- right shift 16-bits
    wop <= shift_right(wop,16);
    zeros(1) <= nand_bits(unsigned(wop(15 downto 0)));
    ones(1) <= or_bits(unsigned(wop(15 downto 0)));
    state <= RT_SHFT_32;
when RT_SHFT_32 => -- right shift 16-bits two times
    wop <= shift_right(wop,16);
    zeros(2) <= nand_bits(unsigned(wop(15 downto 0)));
    ones(2) <= or_bits(unsigned(wop(15 downto 0)));
    state <= RT_SHFT_32_2;
when RT_SHFT_32_2 => -- right shift 16-bits
    wop <= shift_right(wop,16);
    zeros(3) <= nand_bits(unsigned(wop(15 downto 0)));
    ones(3) <= or_bits(unsigned(wop(15 downto 0)));
    -- decide on next step
    if(shifts(4) = '1') then
        state <= RT_SHFT_16;
    elsif(shifts(3) = '1') then
        state <= RT_SHFT_8;
    elsif(shifts(2) = '1') then
        state <= RT_SHFT_4;
    elsif(shifts(1) = '1') then
        state <= RT_SHFT_2;
    elsif(shifts(0) = '1') then
        state <= RT_SHFT_1;
    else
        state <= FINISHED;
    end if;
when RT_SHFT_16 => -- right shift 16-bits
    wop <= shift_right(wop,16);
    zeros(4) <= nand_bits(unsigned(wop(15 downto 0)));
    ones(4) <= or_bits(unsigned(wop(15 downto 0)));
    -- decide on next step
```

```vhdl
        if(shifts(3) = '1') then
            state <= RT_SHFT_8;
        elsif(shifts(2) = '1') then
            state <= RT_SHFT_4;
        elsif(shifts(1) = '1') then
            state <= RT_SHFT_2;
        elsif(shifts(0) = '1') then
            state <= RT_SHFT_1;
        else
            state <= FINISHED;
        end if;
    when RT_SHFT_8 => -- right shift 8-bits
        wop <= shift_right(wop,8);
        zeros(5) <= nand_bits(unsigned(wop(7 downto 0)));
        ones(5) <= or_bits(unsigned(wop(7 downto 0)));
        -- decide on next step
        if(shifts(2) = '1') then
            state <= RT_SHFT_4;
        elsif(shifts(1) = '1') then
            state <= RT_SHFT_2;
        elsif(shifts(0) = '1') then
            state <= RT_SHFT_1;
        else
            state <= FINISHED;
        end if;
    when RT_SHFT_4 => -- right shift 4-bits
        wop <= shift_right(wop,4);
        zeros(6) <= nand_bits(unsigned(wop(3 downto 0)));
        ones(6) <= or_bits(unsigned(wop(3 downto 0)));
        -- decide on next step
        if(shifts(1) = '1') then
            state <= RT_SHFT_2;
        elsif(shifts(0) = '1') then
            state <= RT_SHFT_1;
        else
            state <= FINISHED;
        end if;
    when RT_SHFT_2 => -- right shift 2-bits
        wop <= shift_right(wop,2);
        zeros(7) <= nand_bits(unsigned(wop(1 downto 0)));
        ones(7) <= or_bits(unsigned(wop(1 downto 0)));
        -- decide on next step
        if(shifts(0) = '1') then
            state <= RT_SHFT_1;
```

```vhdl
                else
                    state <= FINISHED;
                end if;
            when RT_SHFT_1 => -- right shift 1-bits
                wop <= shift_right(wop,1);
                zeros(8) <= not wop(0);
                ones(8) <= wop(0);
                state <= FINISHED;
            when FINISHED =>
                -- OR the individual zero and one bits
                dis_zeros <= or_bits(zeros);
                dis_ones <= or_bits(ones);
                rdy <= '1';
                op_out <= wop;
                state <= INPUT_OP;

            when others =>
                state <= RESET;
        end case;

    end if;

end process;

-- gated flags here
op_rdy <= rdy when op_wr = '0' else '0';

end RTL;
```

--
-- csla.vhd Carry Select Adder module
--
-- Multi-stage carry select adder. Number of stages is based
-- on the operand size and the number of partitions needed
-- to support the operand size.
----
library IEEE;
use IEEE.std_logic_1164.all;
use IEEE.numeric_std.all;

entity csla is
-- operand size must be greater or equal to partition size
generic (SIZ,PRTN_SIZ: integer);
port
(
    clk: in std_logic; -- system clock
    rst: in std_logic; -- system reset (must be synchronous)
    -- inputs
    wr: in std_logic;
    ci: in std_logic;
    x: in unsigned(SIZ-1 downto 0);
    y: in unsigned(SIZ-1 downto 0);
    -- outputs
    rdy: out std_logic;
    so: out unsigned(SIZ-1 downto 0);
    co: out std_logic
);
end csla;

architecture RTL of csla is

-- functions
function gen_total_seg_reg(extra,base: integer) return integer is
variable cnt: integer := base;
begin
    if(extra > 0) then
        cnt := cnt + 1;
    end if;
    return(cnt);
end function;

```vhdl
-- data segmenting
type seg_type is array (integer range <>) of unsigned(PRTN_SIZ-1 downto 0);

-- sizing constants
constant EXTRA_BITS: integer := SIZ rem PRTN_SIZ;
constant NUM_SEG_REGS: integer := SIZ / PRTN_SIZ;
constant TOTAL_SEG_REG: integer :=
    gen_total_seg_reg(EXTRA_BITS,NUM_SEG_REGS);

-- local signals
signal a,b,s: unsigned(x'length-1 downto 0) := (others=>'0');
signal c,c_reg: unsigned(TOTAL_SEG_REG-1 downto 0) := (others=>'0');
signal rdy_dly: unsigned(TOTAL_SEG_REG-1 downto 0) := (others=>'0');

-- test signals
signal s0,s1: seg_type(1 to NUM_SEG_REGS-1) := (others=>(others=>'0'));
signal c0,c1: unsigned(1 to NUM_SEG_REGS-1) := (others=>'0');

--------------------------------- module code ----------------------------
begin

-- register process
process(rst,clk)
begin

    if(rst='1') then
        a <= (others=>'0');
        b <= (others=>'0');
        c_reg <= (others=>'0');
        s0 <= (others=>'0');
        rdy_dly <= (others=>'1');
    elsif rising_edge(clk) then
        -- input operands
        if(wr = '1') then
            a <= x;
            b <= y;
        end if;
        -- register selected sums and carries
        c_reg <= c;
        s0 <= s;
        -- ready delay logic
        if(wr = '1') then
            rdy_dly <= (others=>'0');
        else
```

```vhdl
            rdy_dly <= rdy_dly(rdy_dly'high-1 downto 0)&'1';
        end if;
    end if;

end process;
-- last carry in chain is output
co <= c_reg(c_reg'high);
rdy <= rdy_dly(rdy_dly'high) when wr = '0' else '0';

-- base segment adder
base_seg:entity work.cpa generic map(PRTN_SIZ)
port map
(   ci, -- carry in from parent module
    a(PRTN_SIZ-1 downto 0),
    b(PRTN_SIZ-1 downto 0),
    s(PRTN_SIZ-1 downto 0),
    c(0)
);

-- adders for upper segments (minus base segment)
gen_upr_seg: if NUM_SEG_REGS > 1 generate
--signal s0,s1: seg_type(1 to NUM_SEG_REGS-1) := (others=>(others=>'0'));
--signal c0,c1: unsigned(1 to NUM_SEG_REGS-1) := (others=>'0');
begin

    -- create adders and interconnecting muxes
    gen_upr_adr: for i in 1 to NUM_SEG_REGS-1 generate
        -- compute segment for each carry input
        upr_cpa_0:entity work.cpa generic map(PRTN_SIZ)
        port map
        (   '0',-- carry in is = 0
            a((PRTN_SIZ*i)+PRTN_SIZ-1 downto (PRTN_SIZ*i)),
            b((PRTN_SIZ*i)+PRTN_SIZ-1 downto (PRTN_SIZ*i)),
            s0(i),c0(i)
        );
        upr_cpa_1:entity work.cpa generic map(PRTN_SIZ)
        port map
        (   '1',-- carry in is = 1
            a((PRTN_SIZ*i)+PRTN_SIZ-1 downto (PRTN_SIZ*i)),
            b((PRTN_SIZ*i)+PRTN_SIZ-1 downto (PRTN_SIZ*i)),
            s1(i),c1(i)
        );
        -- select sum and carries for current segment, based on previous carry
        c(i) <= c0(i) when c_reg(i-1) = '0' else c1(i);
```

```vhdl
            s((PRTN_SIZ*i)+PRTN_SIZ-1 downto (PRTN_SIZ*i)) <=
                s0(i) when c_reg(i-1) = '0' else s1(i);
    end generate;

end generate;

-- extra-bits adder
gen_xtr_bits: if EXTRA_BITS > 0 generate
signal s0,s1: unsigned(EXTRA_BITS-1 downto 0) := (others=>'0');
signal c0,c1: std_logic := '0';
begin

    xtr_cpa_0:entity work.cpa generic map(EXTRA_BITS)
    port map
    (   '0',-- carry in is = 0
        a(a'high downto (PRTN_SIZ*NUM_SEG_REGS)),
        b(b'high downto (PRTN_SIZ*NUM_SEG_REGS)),
        s0,c0
    );
    xtr_cpa_1:entity work.cpa generic map(EXTRA_BITS)
    port map
    (   '1',-- carry in is = 1
        a(a'high downto (PRTN_SIZ*NUM_SEG_REGS)),
        b(b'high downto (PRTN_SIZ*NUM_SEG_REGS)),
        s1,c1
    );
    -- select sum and carries for current segment
    c(c'high) <= c0 when c_reg(c_reg'high-1) = '0' else c1;
    s(s'high downto s'high-(EXTRA_BITS-1)) <=
        s0 when c_reg(c_reg'high-1) = '0' else s1;

end generate;

end RTL;
```

```vhdl
------------------------------------------------------------
--
--   csa.vhd Carry Save Adder
--
--   Three operand adder. Final carry bit is not generated.
--
------------------------------------------------------------
library IEEE;
use IEEE.std_logic_1164.all;
use IEEE.numeric_std.all;

entity csa is
generic (SIZ: integer);
port
(
    -- inputs
    x: in unsigned(SIZ-1 downto 0);
    y: in unsigned(SIZ-1 downto 0);
    z: in unsigned(SIZ-1 downto 0);
    -- outputs
    s: out unsigned(SIZ-1 downto 0);
    c: out unsigned(SIZ-1 downto 0)
);
end csa;

architecture RTL of csa is

begin

--   carry save adder
gen_csa: for i in 0 to SIZ-1 generate
begin
    -- compute sum out
    s(i) <= (y(i) xor z(i)) xor x(i);
    -- compute carry out
    c(i) <= (y(i) and x(i)) or (z(i) and x(i)) or (y(i) and z(i));

end generate;

end RTL;
```

```vhdl
----------------------------------------------------------
--
--   cs_pkg.vhd (Carry-Save Package)
--
----------------------------------------------------------
library IEEE;
use IEEE.std_logic_1164.all;
use IEEE.numeric_std.all;

--  package header
package cs_pkg is

    -- functions
    function CS_to_T2sC(s,c:unsigned) return unsigned;

end;

--  package body
package body cs_pkg is

    -- converts a number in carry-save form to two's complement, using a
    -- standard carry propagation adder. (for simulation)
    function CS_to_T2sC(s,c:unsigned) return unsigned is
    variable sum: unsigned(s'high downto s'low) := (others=>'0');
    begin

        sum := s + (c&'0'); -- assumes msb of c was discarded

        return(sum);

    end function;

end;
```

```vhdl
----------------------------------------------------------
--
-- cpa.vhd Carry Propagation Adder
--
----------------------------------------------------------
library IEEE;
use IEEE.std_logic_1164.all;
use IEEE.numeric_std.all;

entity cpa is
generic (SIZ: integer);
port
(
    -- inputs
    ci: in std_logic;
    a: in unsigned(SIZ-1 downto 0);
    b: in unsigned(SIZ-1 downto 0);
    -- outputs
    so: out unsigned(SIZ-1 downto 0);
    co: out std_logic
);
end cpa;

architecture RTL of cpa is

signal c: unsigned(SIZ-1 downto 0) := (others=>'0');

begin

-- created with full adders
gen_adr: for i in 0 to SIZ-1 generate

    gen_lsb: if i = 0 generate
        so(i) <= a(i) xor b(i) xor ci;
        c(i) <= (a(i) and ci) or (b(i) and ci) or (a(i) and b(i));
    end generate;

    gen_other: if i > 0 generate
        so(i) <= a(i) xor b(i) xor c(i-1);
        c(i) <= (a(i) and c(i-1)) or (b(i) and c(i-1)) or (a(i) and b(i));
    end generate;

end generate;
```

co <= c(c'high);

end RTL;

```vhdl
--------------------------------------------------------------
--
--  cdr.vhd Cycle Decode ROM
--
--------------------------------------------------------------
library IEEE;
use IEEE.std_logic_1164.all;
use IEEE.numeric_std.all;
use work.sd2_pkg.all;

entity cdr is
port
(
    clk: in std_logic;
    -- inputs
    s: in unsigned(3 downto 0); -- sum bits
    c: in unsigned(3 downto 0); -- carry bits
    divisor_sign: in std_logic; -- sign of divisor
    override: in unsigned(1 downto 0); -- override type of cycle
    -- outputs
    divisor_sel: out unsigned(1 downto 0); -- divisor select
    cin: out unsigned(1 downto 0); -- cin control
    q: out unsigned(1 downto 0) -- quotient SD value
);
end cdr;

architecture RTL of cdr is

type rom_type is array (0 to (2**11)-1) of unsigned(5 downto 0);

-- sets content for 2048 by 6-bits
function InitROM return rom_type is
variable rom: rom_type := (others=>(others=>'0'));
variable adx: unsigned(7 downto 0) := (others=>'0');
variable eval: signed(3 downto 0) := (others=>'0');
begin

    --  (+) divisor, address range 0 to 255
    for sum in 0 to 15 loop
        for carry in 0 to 15 loop
            eval := to_signed(sum + carry,4); -- compute actual value of 2R
            adx := to_unsigned(sum,4)&to_unsigned(carry,4);
            -- (> or = 0)
            if(eval(3) = '0') then
```

```
            -- Q = {+1}, divisor sel = "01" and 2x/cin0, R = 2R - Divisor
            rom(to_integer(adx)) := SD_P&"01"&"01";
        -- (> or = -1/2)
        elsif(eval(3 downto 1) = "111") then
            -- Q = {0}, divisor sel = "00", R = 2R + 0
            rom(to_integer(adx)) := SD_Z&"00"&"00";
        -- (< -1/2)
        else
            -- Q = {-1}, divisor sel = "10", R = 2R + Divisor
            rom(to_integer(adx)) := SD_N&"00"&"10";
        end if;
    end loop;
end loop;

--  (-) divisor, address range 256 to 511
for sum in 0 to 15 loop
    for carry in 0 to 15 loop
        eval := to_signed(sum + carry,4); -- compute actual value of 2R
        adx := to_unsigned(sum,4)&to_unsigned(carry,4);
        -- (> or = 0)
        if(eval(3) = '0') then
            -- Q = {-1}, divisor sel = "10", R = 2R + Divisor
            rom(to_integer(adx) + 256) := SD_N&"00"&"10";
        -- (> or = -1/2)
        elsif(eval(3 downto 1) = "111") then
            -- Q = {0}, divisor sel = "00", R = 2R + 0
            rom(to_integer(adx) + 256) := SD_Z&"00"&"00";
        -- (< -1/2)
        else
            -- Q = {+1}, divisor sel = "01" and 2x/cin0, R = 2R - Divisor
            rom(to_integer(adx) + 256) := SD_P&"01"&"01";
        end if;
    end loop;
end loop;

--  override "01", address range 512 to 1023
for i in 0 to 511 loop
    -- divisor sel = "01", R = R - 2D, cin1/0
    rom(i + 512) := "00"&"10"&"01";
end loop;

--  override "10", address range 1024 to 1535
for i in 0 to 511 loop
    -- divisor sel = "10", R = R + 2D
```

```vhdl
            rom(i + 1024) := "00"&"00"&"10";
        end loop;

        -- override "11", address range 1536 to 2047
        for i in 0 to 511 loop
            -- divisor sel = "00", R = R + 0
            rom(i + 1536) := "00"&"00"&"00";
        end loop;

        return(rom);

    end function;

    -- constant
    constant ROM: rom_type := InitROM; -- initialize ROM based on function

    -- signals
    signal adr: unsigned(10 downto 0) := (others=>'0');
    signal iadr: unsigned(10 downto 0) := (others=>'0');
    signal do: unsigned(5 downto 0) := (others=>'0');

    ---------------------------------- module code ----------------------------
begin
    -- create address from constituents of input signals
    adr <= override&divisor_sign&s&c;

    -- inferred synchronous RAM (ROM), read-through
    process(clk)
    begin

        if rising_edge(clk) then
            iadr <= adr; -- latch address
        end if;

    end process;

    -- propagate through lookup table to outputs
    do <= ROM(to_integer(iadr));

    -- parse out ROM data to output signals
    divisor_sel <= do(1 downto 0); -- selects for +/- and zero cycles
    cin <= do(3 downto 2); -- adds cin and creates 2x
    q <= do(5 downto 4); -- signed digit for quotient
end RTL;
```

8 Other information on Dividers

Other dividers schemes were considered for this book, but were not done because they likely would not be used by the reader. Array dividers use a massive amount of gate resources, and would only be effective if carry propagated downward as in carry-save form. This type of implementation is only practical when designing high-speed ASIC devices, not FPGAs.

With that said, there is ongoing activity in creating faster adders. Some impending FET switches for sums and carries instead of logic gates, some are employing global carry constructs like segmented adders, and so on.

Names of individuals doing work on high performance arithmetic, including division:

    Ercegovac and Lang
    Robertson
    Newtoin-Raphson
    Goldschmidt
    Sweeny, Robertson, and Tocher (SRT)
    Koren
    Cavanagh
    Harris-Weste
    Mark Horowitz

9 Addendum

The source code that follows supports conversions between real and fixed-point number, which exceed 32-bit limit.

```vhdl
--
-- File: ConversionPackage.vhd
--
library IEEE;
use ieee.std_logic_1164.all;
use ieee.numeric_std.all;
use work.all;

-- package header
package ConversionPackage is

    function RealToQmn(r: real; m,n: integer) return signed;
    function QmnToReal(qmn: signed; m,n: integer) return real;

end;

-- package body
package body ConversionPackage is

    --
    -- Convert real number to Qmn fixed-point number
    --
    -- m - number of integer bits excluding the sign (no limit)
    -- n - number of fractional bits (no limit)
    --
    -- Note: an inherent limit to this design is the maximum
    -- negative number is limited to the maximum positive
    -- number in magnitude. This is because the m,n parameters
    -- dictate bit width.
    --
    function RealToQmn(r: real; m,n: integer) return signed is
    variable rx,tmp: real := 0.0;
    variable int: unsigned(m-1 downto 0) := (others=>'0');
    variable frac: unsigned(n-1 downto 0) := (others=>'0');
    variable qmn: signed((m+n) downto 0) := (others=>'0');
    begin
        -- convert to positive
```

```
if(r < 0.0) then
    rx := (r * (-1.0));
else
    rx := r;
end if;
-- compute integer portion
for i in m-1 downto 0 loop
    -- integer size limit
    if(i < 31) then
        -- subtract binary equivalent
        if((rx - real(2**i)) >= 0.0) then
            rx := rx - real(2**i);
            int(i) := '1';
        end if;
    -- multiply up limit by 2
    else
        tmp := real(2**30);
        for j in (m-1) downto 31 loop
            tmp := tmp * 2.0;
        end loop;
        -- subtract binary equivalent
        if((rx - tmp) >= 0.0) then
            rx := rx - tmp;
            int(i) := '1';
        end if;
    end if;
end loop;
-- compute fractional portion (remaining in rx)
for i in n-1 downto 0 loop
    -- integer size limit
    if((n-i) < 31) then
        -- subtract reciprocal
        tmp := (1.0/real(2**(n-i)));
        if((rx - tmp) >= 0.0) then
            rx := rx - tmp;
            frac(i) := '1';
        end if;
    -- divide down limit by 2
    else
        tmp := 1.0/real(2**30);
        for j in 31 to (n-i) loop
            tmp := tmp / 2.0;
        end loop;
        -- subtract reciprocal
```

```
              if((rx - tmp) >= 0.0) then
                 rx := rx - tmp;
                 frac(i) := '1';
              end if;
           end if;
        end loop;
        -- construct final fixed-point number
        qmn := signed('0'&int&frac);
        if(r < 0.0) then
           qmn := (not qmn) + 1;
        end if;

        return(qmn);

    end function;
    --
    -- Convert Qmn fixed-point number to real
    --
    -- m - number of integer bits excluding the sign (no limit)
    -- n - number of fractional bits (no limit)
    --
    function QmnToReal(qmn: signed; m,n: integer) return real is
    variable qmnx: signed(qmn'high downto qmn'low) := (others=>'0');
    variable r,tmp: real := 0.0;
    begin
        -- convert to positive number
        if(qmn(qmn'high) = '1') then
           qmnx := (not qmn) + 1;
        else
           qmnx := qmn;
        end if;
        -- compute integer portion
        if(m > 0) then
           -- add corresponding power of 2
           for i in qmnx'high-m to qmnx'high-1 loop
              if(qmnx(i) = '1') then
                 -- integer size limit
                 if((i-n) < 31) then
                    r := r + real(2**(i-n));
                 -- multiply up limit by 2
                 else
                    tmp := real(2**30);
                    for j in (i-n) downto 31 loop
```

```vhdl
                    tmp := tmp * 2.0;
                end loop;
                r := r + tmp;
            end if;
        end if;
    end loop;
end if;
-- compute fractional portion
if(n > 0) then
    -- add corresponding power of two reciprocal
    for i in qmnx'high-1-m downto qmnx'low loop
        if(qmnx(i) = '1') then
            -- integer size limit
            if((n-i) < 31) then
                r := r + (1.0/real(2**(n-i)));
            -- divide down limit by 2
            else
                tmp := 1.0/real(2**30);
                for j in 31 to (n-i) loop
                    tmp := tmp / 2.0;
                end loop;
                r := r + tmp;
            end if;
        end if;
    end loop;
end if;
-- restore sign
if(qmn(qmn'high) = '1') then
    r := (r * (-1.0));
end if;

return(r);

    end function;

end;
```

www.ingramcontent.com/pod-product-compliance
Lightning Source LLC
Chambersburg PA
CBHW051708170526
45167CB00002B/577